Informed Architecture

Marco Hemmerling · Luigi Cocchiarella
Editors

Informed Architecture

Computational Strategies in Architectural Design

Springer

Editors
Marco Hemmerling
Computational Design in Architecture, Faculty of Architecture
Cologne University of Applied Sciences
Cologne
Germany

Luigi Cocchiarella
Department of Architecture and Urban Studies (DASTU)
Politecnico di Milano
Milan
Italy

and

School of Architecture Urban Planning Construction Engineering (AUIC)
Politecnico di Milano
Milan
Italy

ISBN 978-3-319-53134-2 ISBN 978-3-319-53135-9 (eBook)
DOI 10.1007/978-3-319-53135-9

Library of Congress Control Number: 2017943199

Printed on acid-free paper

This Springer imprint is published by Springer Nature
The registered company is Springer International Publishing AG
The registered company address is: Gewerbestrasse 11, 6330 Cham, Switzerland

Forewords

Rethinking Architectural Education

"Computation is the New Black," the title of the seminar that has inspired this book, refers to an approach taken by Prof. Marco Hemmerling together with Prof. Luigi Cocchiarella to inquire further into the issue of the implementation of computation and digital technology in architecture, and the present-day education in architecture. The seminar was held at the School of Architecture Urban Planning Construction Engineering (AUIC) of the Politecnico di Milano in January 2016, in parallel with an exhibition showing the educational results of two courses focusing on computational design and advanced fabrication, led by Prof. Marco Hemmerling, who was teaching at our School as a visiting professor.

Given the aim of Profs. Hemmerling and Cocchiarella to elucidate the role of digital technologies in architectural education, it is first necessary to define a set of disciplines relevant to this topic, such as architecture and design, representation, mathematics, and geometry. These different disciplines will come together to collaborate in defining tools and the methods of using them in architecture and design.

Our School has become very sensitive to the issue of modernizing education in architecture, in terms of not only tools and strategies but also infrastructure. On the first point, several committees have been at work redesigning our Bachelor's and Master's curricula, and the contributions of visiting professors and international experts from all over the world have been increased, in conjunction with a dense program of conference and seminars promoted and held at our headquarters. On the second point, we have started working on a series of projects, together with the rectorate and the departments, focusing on updating and enlarging our technical laboratories, whose contribution is now even more integrated into the daily educational activities.

The experience which students under Prof. Marco Hemmerling have obtained from studies and research is important in testing this topic and has led to a strong debate in our School at the Politecnico di Milano.

Promoting innovation also offers an opportunity to gain benefits from research, which at the Politecnico di Milano involves several engineering and design fields strongly related to architecture. Nowadays, the computational approach offers a common language that can enormously ease and improve the dialog among experts, with positive effects on not only education but also research and professional practice, as well as their interaction.

Considering the international scenario, that is, the multiplicity of backgrounds of the incoming students, and the complexity of the professional areas related to architecture, an important part of the mission of the School is also to provide adequate support for new students in order to ensure that they are sufficiently advanced to profit from attending the programs, and at the same time to enhance targeted traineeships and work experiences as a part of the educational project.

Accordingly, the School organized a group of professors who have the goal and mission to explore in depth the topic of how to derive tools that can be given to our students in order to enhance the new approach, and to investigate how it can contribute to the present-day architecture, to the students, and to society.

March 2017

Ilaria Valente
Dean
School of Architecture Urban Planning
Construction Engineering (AUIC),
Politecnico di Milano
Milan, Italy

Computation is the New Black

This book tackles a key issue for the theory and practice of architecture: the changes in the design process and production due to the irruption of extraordinary innovations in the field of information technology.

The technical possibilities that new information and computational technologies offer design as well as building processes have consequences that are not only technical. Thinking, drawing, and imagining are deeply influenced by techniques that determine new ways of defining things and their relations.

In this book, the introduction of computational methods in architecture and the use of digital architectural representations are analyzed and interpreted not only as new design tools but also as cultural and social challenges.

The example of the success of Building Information Modelling (BIM) can be useful for further considerations. The power of algebraic and geometric computerized algorithms allows us to design buildings in a completely different way from the past. For example, the ability to foresee the building in a three-dimensional space from the outset contrasts with the traditional design process involving initial description through plans and sections (and therefore in two dimensions) and implies what can be called a new "writing" of the architecture. However, BIM is also a tool for control of the entire project chain, from concept to construction site, within a unified program. It represents a radical revolution for architectural design, which depends on the one hand on the computational power of information technologies and on the other hand on the power of representation of computer graphics programs.

If we consider the importance of tools and instruments in design practices, we should then be able to answer many different questions:

How do specific and concrete design practices happen through the use of these new "tools," which are still instruments (such as the pencil, the paper, and the drafting machine) but which "write" the world in another way?

What are the reasons why these new tools are being deployed? There are aesthetic reasons, connected with the possibility of generating new and different forms, but in addition crucial economic and organizational reasons have to be taken into account. BIM optimizes the building production process, fully integrating the project activity into construction practices and, through this integration, redefining the social and technical role of the architect.

What consequences will these techniques for design thinking and their outcomes have, based on the way architects and architecture students practice the project, for the necessary innovation in our academic strategies and training programs?

All these serious questions need a profound understanding of the opportunities, challenges, and risks that this revolution will bring in the coming years. This book offers a fresh and articulate representation of these opportunities and challenges, in an interdisciplinary perspective that takes into account both the theoretical issues raised by computational techniques and the economic, social, and cultural dimensions of this reorganization of the entire sequence of architectural production.

March 2017

Gabriele Pasqui
Head
Department of Architecture and Urban Studies (DASTU)
Politecnico di Milano
Milan, Italy

Preface

The future is above all a question of design (Vilém Flusser).

The impact of digital technologies is perceptible on many levels, and since the process of implementation of such technologies is still in progress, it will also determine our future. Predicting the future has always been a dream of man. Nevertheless, it is still the case, perhaps fortunately, that the future remains unpredictable. Over recent years, however, computational tools and methods in architectural design and construction have developed rapidly and now allow for an approximation and simulation of the future in our profession. At the same time, complexity is increasing and the frontiers between the professions are becoming more and more permeable. This is both a big opportunity and a demanding challenge as new concepts and promising solutions can only be achieved by interdisciplinary work, where each field contributes its expertise, methods, and different points of view. Managing this process requires a holistic approach rather than focusing on individual aspects. In other words, we are looking for strategies of integration and a process-oriented perspective in architecture.

Information technology has brought about lasting changes in design and production processes in architecture. At the same time, our demands in respect of design and building processes have increased in line with technical possibilities. Aside from unprecedented geometrical freedom and new fabrication technologies, there is huge potential to optimize functions, energy usage, and performance of constructions, buildings, and services.

The programmatic title "Informed Architecture" connects the different topics and professions involved from a holistic perspective, ranging from Computer-Aided Design to Building Information Modelling, from Programming to Simulation, from Digital Representation to Augmented and Virtual Reality, and from Digital Fabrication to Physical Computation. In this book, experts from these fields contribute their academic and practical experience and their findings in research and advanced applications. The interdisciplinary contributions to this book cover the fields of architecture, engineering, design, and mathematics. In addition to these scientific papers, documentation of academic projects illustrates architectural case studies that were carried out at the East Westphalia University of Applied Sciences and the Politecnico di Milano. Downloadable interactive digital graphics samples related to the mentioned projects will be included as supplementary information.

Against this background, the publication not only showcases the broad range and impact of information technology in architecture from an academic point of view, but also discusses different teaching methods and future developments in the field.

As such, this book will serve as an inspirational source for students and lecturers in architecture, design, and engineering as well as a state-of-the-art overview for researchers and professionals. Moreover, the volume will contribute to interdisciplinary discourse and aims to foster the critical discussion about the opportunities and risks for our profession in the digital era.

What are the conditions, constraints, and opportunities of this digital turn for the conception and making of Architecture? How do processes change and influence the result? What does it mean for the collaboration and roles of the partners involved? And last but not least: How does academia reflect and shape this development and what will come next? Following the sequence of architectural production—from design to fabrication and construction and the operation of buildings—the publication discusses the impact of computational methods and technologies and their consequences for the education of future architects and designers. Hence, this book aims at an in-depth understanding of the processes involved and reflects them in respect of our technical, historical, social, and cultural environment.

Prospective interested readers of the volume are all those academic and professional operators involved in the above-mentioned fields, including Bachelor's, Master's, and Ph.D. students, to whom this work could be proposed as a textbook, that is, as a theoretical as well as an operational reference.

Milan, Italy
March 2017

Marco Hemmerling
Luigi Cocchiarella

Acknowledgements

This publication was made possible with the support of the School of Architecture, Engineering and Urban Planning as well as the Department of Architecture and Urban Studies at the Politecnico di Milano. Special thanks go to Freeda Jane Madius and Julia Popp for their supportive work on the text editing and layout of this book. Furthermore, we would like to thank all the assistants and students involved in the academic projects for their ambition, effort, and relevant contribution to this publication.

Contents

Part I
Information Technology in Architecture

Informed Architecture

Marco Hemmerling

New types of space need new styles of architecture.
(*Marcos Novak* [1]).

1 Introduction

Regardless of their content, new technologies cause changes in perception and thinking. They create new realities or, as the Canadian media theorist Marshall McLuhan [2] puts it: "We shape our tools and then our tools shape us". For McLuhan, the development of media technologies thereby acts as a driving force for social change. Digital content influences the spaces in which we live, the objects that surround us, the images we see, and the sounds we hear. In this way, they further our insight and perception of our reality. This development has resulted not only in a radical transformation of our environment, but has also created a space in which architects can design and build. The Digital is becoming a catalyst for a new architectural vision.

Digital Technologies have manifold implications on different levels of the architectural process. Computational design strategies, digital fabrication methods and virtual applications lead to a significant acceleration in the generation, production and appearance of architecture. Technology always had a major influence on the profession of architects, engineers, and designers. Like Le Corbusier [3] stated in his manifesto "Toward an Architecture", technology was not only a formal

This article is based on the following publications that the author published in 2011: *Die Erweiterung der Realität* in: *Augmented Reality*—Menschen, Raum und Virtualität. Hemmerling, M. (Ed.) pgs 7–24, Wilhelm Hink Verlag.München.

M. Hemmerling (✉)
Cologne University of Applied Sciences, Cologne, Germany
e-mail: marco.hemmerling@th-koeln.de

M. Hemmerling and L. Cocchiarella (eds.), *Informed Architecture*,
DOI 10.1007/978-3-319-53135-9_1

inspiration for a new architectural style, but altered also the entire architectural process from design to construction and operation of a building. Nowadays, advanced technologies influence our profession again in a revolutionary way. The use of contemporary computer-aided tools has not only significantly altered the way architects work, but has also greatly influenced the formal design process and the resulting appearance and perception of space.

Using computer software, designs can be generated in ways that until now have scarcely been possible, technically and formally. To a large extent, computer technology frees architectural design from the traditional conditions of production. Dependencies are shifting, both in the design and the production process, from analogue to digital processes. The computer expanded the boundaries of the imagination and allowed forms to be generated that would formerly have failed because of limited time and technical resources. The results of this fascination with what is digitally feasible, however, frequently goes no further than experiments with form, often alarmingly one-dimensional despite their geometric complexity. Greg Lynn, one of the proponents of free digital worlds of forms, gave a negative judgement of this formal dependency as early as ten years ago: "*There is a language of design that the computer brings with it, and mainly, you do what the software does well*" [1]. The potential of computer-aided design and construction is far more promising than this quote suggests. However, it requires an understanding of the basics of information technology that the computer provides. The result of this kind of architectural production is followed by other requirements and constraints. It is based on a new methodical understanding, both on the level of design and in realization. But what does design look like in the digital age? What developments are on the horizon and what influence do these technologies have on architecture and the role of architects in the future?

2 Extension of the Combat Zone

At a basic level, we can say that the introduction of information technology has led to a significant extension of the professional field of architecture. New and interdisciplinary fields of activity are opening up to the architects who are open to new technologies. In addition to changes in design methods brought about by new aspects of architectural creation, the design skills of the architect can be transferred to many areas of media and information technology as well as construction technology.

The influence of digital media in the design and realization process on architecture can be derived from two evolutionary strands. On one side, the computer supports the creation of architectural concepts using digital tools. The complete representation of three-dimensional design concepts using digital simulations and direct interaction with the virtual model in the design process substantially extends the perception of spatial relationships. In this way, computer-generated representations mean that spatial concepts can be seen, evaluated, and communicated early in the design process and in their entirety. From the first design sketch and concept visualization, to the final 3D building data set, the computer now maps the entire design and planning process [4]. Nevertheless, virtual worlds, such as those used in computer games, are still seen as independent areas of design that seem largely inaccessible to architects. The design of spaces in this field is based on similar approaches and tools for reality and virtuality. In particular, the observation of a virtual building or room design in real-time from a first-person perspective is an important asset for the evaluation of architectural concepts [5]. The goal of the use of virtual reality in practice, therefore, is to approach an architecture whose effect, and as many of its consequences as possible, are known and experienced before its execution.

On the other hand, in recent years there has been a sharp increase in computer-aided manufacturing processes using *CAD-CAM* interfaces and *rapid prototyping* processes. What is special about this technology is that different shapes can be produced using one and the same manufacturing process. As this operation is fully automated, the price of production is, in principle, the same. A new freedom in design for architectural production results from this since the concern of cost is largely removed from the standard product. The manufacture of individual products using industrial methods of production allows design—and customer—specific mass production in architecture, as has already been established in other areas of production. A look at the main areas of application for digital tools makes it clear how specialized computer-aided methods already have an influence on the development and formal expression of architecture.

3 Digital Design

The architectural design process is characterized by an intuitive approach that is seemingly difficult to reproduce on a computer. Few architects use a computer exclusively in the production of their drafts. Nevertheless, particularly among students and the younger generation of architects, we can see the pen being replaced by the computer in the early design phase. But the potential of digital design is not in the simulation of operations that were formerly analogue, but on the harnessing of processes inherent to the computer to capture, link, process, and evaluate complex interrelationships. Here we find an essential difference from traditional CAD applications, which might have supported the drawing process, but did not have an effect on the method of design itself.

New digital tools now allow spatial concepts to be developed and modelled using processes while taking into account various parameters. A basic parametric model first defines the environment before the initial geometry of the design is provided with algorithms, i.e. mathematical instructions, which influence the geometry and other aspects of the model, such as material properties and design parameters. Many software applications today offer the option of programming custom scripts and algorithms, which can be used to define the boundary conditions for the production of the design individually. The architects develop their design tools themselves. This process-oriented method allows the scale-crossing manipulation of the entire structure within the development of the design without losing the linkages between the individual building components among themselves. The design process is supported, to a large extent, by information technology in this kind of architectural production. Programmed parametric models in this sense represent a new methodology of design that simultaneously requires knowledge that is specific both to architecture and to computers.

In relation to specific issues in architecture, parametric models can be used to create adaptable structures that react to outside influences such as sunlight and wind exposure or internal influences such as user behavior and functional processes. The form of the design arises through the definition of linkages and prioritization of individual parameters. Design shifts from a formal and graphical process to a strategic and evolutionary process. This means that the architect designs the process more than the result of that process. An advantage of this approach—apart from the visibility of relationships in the design phase—is the ability to influence the programming at any time and therefore the ability to develop and evaluate different concepts or variations of a design solution.

The principles of parametric design also form the basis for *Building Information Modelling* (BIM).[1] This term refers to a digital planning methodology in which all processes around the design, planning, execution, and management of a building are connected together in a network. In addition to graphical information on building geometry, this includes non-graphical information such as information on quantities and materials, ecological and economic values, and the definition of functional areas and usage profiles. The modifications that are carried out on these levels influence other areas directly, regardless of whether the 3D model, the floor plan or the list of building components are adjusted. As well as a comprehensive link between all building relevant information for architectural planning, BIM also defines the interfaces to other areas of planning: e.g. structural engineering, building physics or *facility management*.

4 Digital Construction

Computer-aided production methods enable the direct translation of this digital 3D data into physical models or components. Digitally generated and mechanically manufactured components extend the constructive spectrum and incorporate the constraints that result from the choice of materials and manufacturing logic into the

[1]http://buildingsmart.org/.

design process. However, these manufacturing methods do not necessarily result in a formal change in architecture. Rather, they help to develop and optimize processes. Therefore, the formal result is initially independent from the process. In this sense, the spread of digital manufacturing should not only be attributed to individual buildings of high complexity. But in comparison to other areas of production, designs in architecture are still implemented largely using traditional working processes. While high-tech products appear in industrial production processes, such as machine construction or the automotive industry, nearly every building that is designed is a prototype that is built using conventional methods. This construction process is not only cost and time intensive, but also does not make use of the creative freedom afforded by digital design development and computer-aided manufacturing. Developments in materials research and construction technology definitely points in a new direction. In the Digital Fabrication—Department of Architecture, ETH Zürich [6], machine-aided processes are developed further and linked to digital design methods. The automated construction of a brick-masonry wall by industrial robots that are controlled by an algorithm allows a multitude of variations within a programmed production process. Insights from manufacturing processes then flow back into the design process as parameters for programming and therefore inform digital and physical architecture.

5 Mediatecture

There is a new profession at the interface between physical spaces and virtual worlds, at the boundary of architecture: Mediatecture. This term, coined by Christoph Kronhagel [7], describes the integration of digital media in a physical/spatial context. Mediatecture therefore occupies the point of intersection between concrete locations and virtual worlds, and links the two in a holistic approach to design. This changes and expands the perception, function, and utility of our constructed environment and new relationships between people and space are created.

In addition to controlling important functions of building management, digital technologies also support communication and interaction between users and architects. The materials and structure of the building are made increasingly accessible to computerized processes and the functions of a house are based more and more frequently on digital interfaces. In such structures, the individual components communicate using various external and internal networks and allow access to a variety of building data. Meanwhile, architecture can be individually and differentially designed, from regulation of temperature and sun protection depending on the weather, to alarm systems and lighting profiles.

The visible points of intersection between space and user are disappearing as more and more structures are outfitted with digital technology. Contactless sensor technology, capacitive surfaces and wireless access to technology all mean that various functions in the room can be controlled virtually invisibly. Automation is increasingly linking individual processes to complex, technical processes inside and outside of the building. The effect of this development in technology on the relationship between people and space is described by media theorist Oliver Grau [8] as follows: "*The higher the apparent resolution of the interface, the more intuitively, more naturally the interface is designed and therefore the illusory symbiosis between the viewer and the work advances, the more the psychological distance recedes*". An example of this development is Wohnhaus R128 by the engineer Werner Sobek in Stuttgart, in which the technology is invisible, but its perception and use are obvious. The dematerialization of the fully-glazed, energy self-sufficient building, which has no interior walls, only further strengthens the presence of the ubiquitous technology. The user makes contact with the invisible but active building technology in many ways. Most operations in the house require communication with the physical architecture with voice control, touch pads or motion sensors. The user's movement becomes a dynamic factor of movement between rooms, which reduces the relevance of physical boundaries between spaces. The increasing integration of technical devices such as smartphones, tablet PCs and laptops into daily routines is intensifying the user's awareness of the effect of the subjective space continuum throughout the built environment.

As a consequence of this development, the significance of building automation is growing as an architecturally relevant aspect compared with the classically physical production of spaces. This expands the field of activity in the direction of a user-dependent variability in spatial perception and thereby creates a new level of interaction between people and spaces. It is more obvious that digital media are making architecture more dynamic when this dynamization is developed and made recognizable as early as the design process as the basis of creative intent. Here, architects and designers are encouraged to understand this as part of their planning object and to incorporate it into their design.

The level of media communication shifts the weighting of the context, which we have known until now as the real environment. The importance of the physical and material environment becomes less as the attractiveness of the informational context increases. In future, the flexible structures of informational architecture will come to have a stronger influence on our view of reality. In this process, the superimposition of real and digital information will create a new and appealing spatial experience. But how can the different methods and technologies be used to develop a sustainable architecture? And how can these be used to define an independent expression of architecture that fulfils contemporary and future demands?

6 Digital Realities

A major advantage in the use of computer-aided methods is the many opportunities to connect individual processes together strategically, to utilize synergies and recognize relationships early on and make them usable. The basis of these process-oriented strategies is the development of a consistent and adaptable design model that is creatively developed further in the planning process and successively elaborated upon and expanded by additional information. The result is an integrated architecture that arises from the interaction of factors such as spatial impact, form finding, materialization, conditions of construction, and production, as well as functional requirements and user behavior.

Digital design creates a direct link between what it thinkable and what is buildable. Furthermore, our design and action space is constantly being expanded by the "information" of architecture. Intelligent components that collect and process information using cyber-physical interfaces and sensors are a part of our built environment. This development was clear to see at the central exhibition "*elements of architecture*" at the Venice Biennale of Architecture in 2014 [9]. Associated questions about the position of architecture should be part of a wider debate. It seems to me that this means we take on a responsibility as designers to make the new technologies usable in our processes. Digital tools and content—from the design stage to realization and beyond—not only strengthens the competence of the architect, but allows us to create living spaces that satisfy the demands of contemporary architecture.

The computer is certainly the most comprehensive and dynamic medium that has ever been available for an architect and their work. Organizing this potential, however, requires the use of the computer as an interactive instrument and an understanding of its artificial intelligence as an extension of our creativity. Architects should be encouraged to fulfil this role in our information society, because the ability to deal with digital media will allow architectural quality to remain within the skill set of the architect and at the same time expands their opportunities to create new spaces.

The first reality of the physical space and the virtual reality of digitally created environments are increasingly merging into an emergent overall experience. The area of activity is expanding and the means of design that are available to us are becoming more diverse. Above all, they create a link between the two worlds. Hani Rashid, co-founder of the New York architectural firm Asymptote [10], describes the design field of augmented reality as follows: "*We will continue to see experiments with the virtual that leave the confines of the screen, and merge the virtual with the real, spaces that will ultimately blur the distinctions of what we currently think constitutes a real experience versus a virtual experience*".

Defining the boundaries between one reality and the other is becoming more and more difficult. Just as partial experiences solidify into a whole in our consciousness, we will begin to think of and design the physical and virtual parts of our lives together. Emergence arises accordingly through the connection of different parts into a new quality of space. This being the task for architects is well-founded in their profession, at the point of intersection between human and space.

References

1. Novak, M., Liquid-, Trans-, Invisible-: The Ascent and Speciation of the Digital in Architecture. In: Cachola-Schmal, P. (ed.) Digital real—Blobmeister. Birkhäuser, Berlin (2002)
2. McLuhan, M.: Die magischen Kanäle - Understanding Media, Düsseldorf, Wien (1970)
3. Le Corbusier.: Toward an Architecture. Translated by John Goodman. Getty Research Institute, Los Angeles (2007)
4. Hemmerling, M., Tiggemann, A.: Digital Design Manual. DOM Publishers, Berlin (2011)
5. Hemmerling, M.: *Let's Play! Computerspiele und die digitale Revolution der Architektur*, pp. 187–200. In: Hofmann, M. L. (ed.) Design im Zeitalter der Geschwindigkeit, Wilhelm Fink Verlag München (2010)
6. Gramazio, F., Kohler, M.: *Digital Materiality*, Lars Müller Publishers (2007)
7. Kronhagel, C.: Mediatektur: Die Gestaltung medial erweiterter Räume. Springer, Wien (2010)
8. Grau, O.: Virtuelle Kunst in Geschichte und Gegenwart, Visuelle Strategien, p. 183. Reimer, Berlin (2001)
9. Koolhaas, R, et al.: Elements, Box Edition, Marsilio Publishers (2014)
10. Rashid, H., Couture, L. A.: Flux. Phaidon Press, London (2002)

The Colors of Black: Digital Computation as a Spectrum of Knowledge

Luigi Cocchiarella

Fig. 1 Integrating computing, feeding knowledge (drawn by the Author)

1 Introduction

The title refers to the seminar *Computation is the new Black. Updating the Making of Architecture* organized by Prof. Marco Hemmerling at the School of Architecture Urban Planning Construction Engineering (AUIC) of the Politecnico di Milano in

L. Cocchiarella (✉)
Department of Architecture and Urban Studies (DASTU), Politecnico Di Milano, Milan, Italy
e-mail: luigi.cocchiarella@polimi.it

L. Cocchiarella
School of Architecture Urban Planning Construction Engineering (AUIC), Politecnico di Milano, Milan, Italy

M. Hemmerling and L. Cocchiarella (eds.), *Informed Architecture*,
DOI 10.1007/978-3-319-53135-9_2

January 2016, and the contents to some issues presented in the public speech prepared for that meeting.

Computation has always been an obsession for architects and not only in the digital era. Throughout long history, formulas, ratios, and golden proportions have worked as helpful tools and sometimes even as exorcisms against the fear of failures in building as well as in investigating architecture. Already acting, since the fight of Abacus masters against the apparent contradiction between Euclidean 'ratios' and perspectival 'cross ratios', computational needs could be also found in the origin of Descriptive Geometry, the first modern operational connector among design, industry, and construction. As Robin Evans argued [7], indeed, this discipline proposed a cross fertilization between analytic (numerical) and synthetic (graphical) approaches, conducting visual representation from the world of Art to the world of Science and Technology. However, only geometrical shapes could be accurately defined and controlled, while strategies and processes remained unrevealed. Black is Black.

The revolution of digit, instead, has offered "neutral" *units* as supportive syllables for any kind of information, manageable through algorithms. Many aspects and categories of parameters could be then incorporated and related in the body of representational models, dynamically developed, and widely shared via network. Therefore, together with spatial data, time, and any other informative dimension could increase the syncretism of the models, paying homage to the Juri Lotman's idea of *semiosphere*. Moreover, they could be analytically recalled or re-organized at any time. Colors of Black would then appear, like those of white in a Newton's light spectrum, that is, a spectrum of knowledge.

2 Visual Geometry as the Trailblazer

Since architectural computation is related to *form recognition, form finding*, and *form definition*, the role of *image* is clear in all the processes normally related to the work of an architect. Moreover, since the basic matter of architecture is *space*, the role of *geometry* is clear in the mentioned processes. Summing up, the role of *geometry of image* arises. The need for a clear understanding of the relationships among *geometry of vision*, *geometry of things*, and *geometry of image*, was from the beginning so strong as to generate a prodigious revolution inside the monolithic body of Geometry, namely, of Euclidean Geometry, that is, the *geometry of things*. Quite paradoxically, the source of discord dates back to Euclid, the same official father of Euclidean Geometry. As soon as he wrote the book of Optics, which opened the way to the investigation of the *geometry of sight*, he revealed indeed a problematic conflict between the perfect measurable world of Euclidean objects and the apparently instable and distorted world of Vision. Consequently, the controversial debate about the *verity of images*, already started some decades before with the philosopher Plato, would have been even more exacerbated [14]. Hard enough to believe, not even *perspective*, the pearl of Renaissance, relieved mathematicians or artists, in spite of the further clarification of the geometric links among objects, vision, and (perspective) image.

What was the still pending question? Just *computation*: although the graphic procedure was clear, as well as the related physical phenomenon of projection/section, the analytical side of the problem of how to calculate the perspective distortion of any set of points was still to be defined. Thanks to the work of some giants of Western science, from Girard Desargues and Blaise Pascal to Jean Victor Poncelet, the problem was not completely codified until the XIX century. The understanding of *cross-ratio* (or *bi-ratio*) as a projective invariant, finally provided a convincing computational solution to the long-existing brain-teaser, but the cost was undoubtedly high: the official birth of Projective Geometry as a separate branch from Euclidean Geometry. By the way, Topology had already shown new geometric ways of thinking of space, by denying the fifth postulate of Euclid, while other geometric branches would have been developed over time. However, despite the strong impact of the new theories, Geometry kept its profound unity, re-defining its foundations on the bases of a wider set of axioms, as Felix Klein pointed out in the *Erlangen Programme*. Back to our purpose, an excellent way to mitigate the friction between real space and image was the use of Orthographic Projections, intuitively shown since the origin of representation and officially codified only at the end of the XVIII century by Gaspard Monge. Saving parallelism and proportions according to some given conditions, and offering the opportunity to generate *bi-jective* representations, this system made metrical computation easier to master, reason why it soon became the *symbolic form* of the First Industrial Revolution, and is still valid nowadays.

Satisfactorily enough for the applications, from a theoretical point of view, this was actually a hybrid solution. Indeed, real world continued to follow the computational principle of *ratio*, while the world of projective image was based on the computational principle of *bi-ratio*. In other words they were brothers, but not twins. Finally, a unitary hypothesis seemed to emerge from the Le Corbusier's treatise *Le Modulor* [9], which is just quickly outlined in (only) one illustration, namely the Fig. 31 in Chap. 4 of the first volume, partially recalled in the Fig. 74 of the Chap. 7. As we know, the two volumes are totally dedicated to the search of golden proportions in the true objects, but in the mentioned image *golden ratio* is also proposed as a criterion for computing optical foreshortenings in the visual perception. Based on this hypothesis, and in relation to Architecture, Euclidean and Projective spaces could be considered as based on the same metrical structure, and the *golden ratio* as the universal computational unit to achieve visualization, configuration, and construction, so reducing the traditional huge "trouble with numbers", as Robin Evans defined the classical dichotomy between metrics of space and metrics of image.

Actually, although fascinating, the *golden ratio* has never become *the* architectural unit in use. Architects basically continued to refer to the hybrid system (half visual-half metric) of the orthographic projections, especially during fabrication and building construction. Wonderful drawings finalized to construction can be found anyway, even before the codification of the Monge's procedure, like those by

Philibert De L'Orme in XVI century [6], as well as those by Amédée François Frézier in XVIII century. A key step towards more complex parameterizations was the use of orthographic views as a base for structural calculation, that is, *graphic statics*, where the invisible geometry of loads was integrated with the visible geometry of space. Extraordinary examples of the claim for this union between *aesthetics* and *statics* are the *tableaux* of the project of Gustave Eiffel's Tower as well as many other projects developed by engineers between the XIX and the XX century. In such examples, drawings and diagrams work together in such a way that changes in one layer imply changes in the others as well. In other words, we can consider such examples as the closest ancestors of parametric design. More difficult was the attempt to introduce computation in the whole architectural design process, since it is based not only on physical issues but also on many other sets of inputs, therefore requiring a wider holistic approach. Great help in this field came from Humanities during the XX century, especially from Semiology, namely with *Structuralism* and *Post-Structuralism*. The focus of this approach was on the identification of elements, relationships, and hierarchies in the observed systems, whether linguistic, social, or anthropologic and so forth. In the middle of the XXI century, *Architectural Structuralism* and *Post-Structuralism* arose, which emphasized the role of structures and hierarchies among structures, in architectural space as well as in design processes. Constructed results like *Nakagin Capsule Tower* by Kisho Kurokawa or *Habitat 67* by Moshe Safdie clearly show this approach. Few years before, a graphic synthesis among space, statics, and music, had been achieved in the *Philips Pavilion* project by Le Corbusier and Iannis Xenakis. Soon this approach expanded further, opening new perspective towards phenomenology, deconstructionism, and so forth, but the basic idea of a structure as representative of a process survived, ready for the decisive marriage with *digital algorithms*, from which present *digital computation* comes.

What is, again, the fascinating breaking news? In our opinion, *image*. Or better yet, *imaging*. That is, the possibility to visually control and manage several sets of parametric structures, testing, matching, and combining them in real time, with a direct feedback in terms of spatial organization and information data. In this sense we see *image* in the history as a *trailblazer* to computational design. What also confirms our identity as "visual animals", as John Barrow clearly pointed out in the book *Cosmic Imagery: Key Images in the History of Science* [1].

3 Tactile as a Visual Extension

Digital Process Design also became as important as (and of course intrinsically related to) Object Design and Construction. Together with definitely encouraging brotherhoods between theory and praxis, it finally freed architects from at least two long lasting nightmares: paradoxes of "representational clones" evoked by Borges, and dreads for "uncanny valley" effects noticed by human-like robot designers. It was clear, indeed, the dominion of architects on the new virtual architectural clones, which are modifiable at any time. Last but not least, the benefits of the dialog between *analog virtuality* and *digital virtuality*, according to the Pierre Lévy's clues about "virtual", soon had increasing effects on the hybridization between tangible physicality and immaterial information inside the built architectural environment, claiming a wide expansion of architecture towards the world of the *internet of things*. Furthermore, let us say, with reference to architects and inhabitants, towards the *internet of bodies*.

Originally, space composition and construction were synesthetic matters for human beings, where any process was based mainly on the coordination between vision and touch, and of course on the other senses. Later on, the need for agile operational testing models pushed human beings to choose graphics as the main supportive tool. Thousands of years have been spent using graphic models to simulate architectural composition and construction, and at least more than two millenniums struggling with the dialogue between geometry of space and geometry of image, as we mentioned before. In the past, physical *sculptural* operations were translated into *pictorial* codes, that is, tracing lines on a sheet. Operations like cutting, moving, drilling, and others, could be only represented as pictures but not as actions. As we know, an analogue *graphic menu* consists of a set of points and lines, together with rules, compasses, and various kinds of templates. Physical modelling, enormously more affine to the real process of construction, has efficiently flanked graphics, but maquette-making has never surpassed drawings. As we often see, in our academic educational work too, physical modelling is quite a demanding activity, and in addition, it mostly requires preparatory geometrical drawings as well. However, sometimes, quick-sketch models are extremely supportive for spatial understanding and orientation. In general terms, compared with the visual and metrical control provided by drawing, physical modelling clarifies *spatial topology*, implying a more direct simulation of the compositional and constructive process. We all know, indeed, how very helpful and irreplaceable the power of tactile manipulation is.

Well, *digital graphics* is where the visual and the tactile powers of representation finally meet, showing image and process at the same time. Paradoxically, toolbars are undoubtedly graphic, showing palettes where physical actions and motions appear sublimated as graphic icons together with purely graphic commands (revenge of painting) but operational devices, including the mouse, work in a properly said three-dimensional space (revenge of sculpture). We see in this *a tactile extension of sight*. It was clear since the beginning, dating back to Ivan Sutherland's *Sketchpad*, that computer graphics was born not merely to clone painting but to finally conquer the space.

4 Data Versus Form Versus Knowledge

After early enthusiasm for the high-performance shape manipulation and rendering offered by digital means, architects started deepening the relationships between architecture and information technology even more. Among the leading examples in Europe, the work of Ludger Hovestadt's research group at the ETH of Zürich could be mentioned. In charge of the CAAD chair since 2000, he drove research on a new track, inspired by the motto "our credo was not *virtual reality* but *back to reality*". The result was a very rich series of analyses and design experiments carried out on various scales and aspects of architectural project and construction. An interesting picture of the start of this work is given in the book *Beyond the Grid—Architecture and Information Technology. Application of a Digital Architectonic* [8], where in spite of the declared non-theoretical ambition, a basic organization of the matter emerges. Starting from investigations on patterns, the team focuses on architectural elements and systems as well as on construction, indicating *global design* as a final goal. Significantly, they liked to define their field of interest in terms of *applied virtuality*. Although at a first listening this definition could sound like an oxymoron, we soon realize that it actually makes sense as one of the most urgent and promising fields of investigation. In a wide sense, in fact, virtualization is the essence of any project as well as of any model of knowledge, independently of its analogue or digital consistency.

The power of algorithm is the main feature in this story, offering extraordinarily refined simulations in search of the best possible architectural solutions. Of course one point would be how to control and dominate this powerful operator. Another key aspect concerns the way we choose and translate natural inputs into appropriate and effective parameters. And finally how all this turns into a buildable architectural shape. Therefore, the need to have various sets of expertise at work is clear, side by side, beyond the machine, and consequently a change of style in the approach to the architectural project. Apparently only based on technical restrains, these experiences claimed from the outset the need for a cultural change.

Parallel to these pioneers and their genuine research, the market started developing products in search of new professional standards. As a result, Building Information Modelling, that is BIM, emerged, where letter M could also stand for Management, as Steve Race states in the well known book *BIM Demystified* [12], meaning how very important collaboration and human control are in this field. In spite of the recurring opinion about BIM, technology is not its only strength. Even more decisive, although apparently hidden, is methodology. As we know, architecture involves many fields, and architectural academic curriculum is one of the most rich of different disciplines, including Humanities as well as Art and Science. This variety implies that an architect has to work as a kind of movie director, where characters are disciplinary specialists and the movie is architecture itself. In the past, this connecting role only depended on the personal abilities of the architects, who collected inputs from everybody, taking them into account and synthesizing them in their projects. Digital systems support the architects better in this nowadays, by recording, combining, integrating, and activating the given sets of data, but the creative freedom of the architect's risks to be inhibited if they are not adequately aware and skilled to manage such complex databases. In the afore mentioned book, Race himself affirms that BIM is, in the end, and we could affirm the same for the whole field of computational design, "a state of mind".

Here is where *Structuralism* shows its heritage and, at the same time, a claim for a close connection with the tradition of *Encyclopaedia*. They both, indeed, deal with the aim of organizing knowledge. Looking at the problem from an academic point of view, it is easy to understand that computational design and BIM can be seen also as organized archives, which though, different from the past, are not static but based on a dynamic and interactive indexing. In this sense, we are nowadays dealing with a new connection among *data*, *form*, and *knowledge* in architecture, engineering, and design.

Consequently, apart from its professional impact, computational design can really inspire a new dynamic taxonomy of knowledge, and BIM can be seen as a modern system for building up the archives of knowledge, suggesting in addition stronger interaction with other areas of interest, first of all GIS, which would require a separate dissertation. And, of course, computational design can offer efficient new bases to education, both as a tool and as a reference paradigm. As we tried to point out in the three volumes *The Visual Language of Technique* [4], specialization and expansion of knowledge from the XVII century onwards has worked in favor of the disciplinary "ivory towers", while the unifying power of digital seems to invite our academic and professional communities to build up "virtual bridges". Considering the present global scale, this will be an urgent point of the future educational strategies, again in the words of Steve Race, aiming at "educating people but also government commissioning teams", together with students, and of course, teachers.

5 Final Gloss

When on this topic, we normally end up asking questions about the future. Concerning expectations, apart from diatribes, like the one superbly undertaken between Nicholas Carr [3] (*The Shallows. What the Internet is Doing to Our Brains*) and Howard Rheingold [13] (*Smart Mobs: the Next Social Revolution*), thanks to the exponential power of the network in the world-wide-web, "process" itself is supposed to be the main topic nowadays, and in a way our main invisible competitor, as prophetically announced by Derrick De Kerckhove in the book *Architettura dell'intelligenza* [5].

The whole result of the trend, as depicted above, and the most welcome as well, would not only be about improving architecture making, but most of all in empowering our imagination, let us say, to push ahead the limits of *architectural utopia*, aiming at leaving a better and healthier architectural environment to next generations. Of course on this huge ocean of information we are required to take the helm and trust, feed, and update our original "cybernetic" aptitude as navigators and explorers, amazingly echoing the ancient Greek verb κυβερναω.

But where is our thumb line in this ocean? In attempting to answer this question, we propose to seek help from Italo Calvino's [2] *Six Memos for the Next Millennium*, where in relation to Literature he said that since the mission is to tell stories, nothing happens without words. So we could observe that, since our mission as architects to build places for human beings and understand places of human beings, no matter how much our powerful and sophisticated tools show, nothing happens in architecture without space. Let's then continue by suggesting that as much as words require grammar and syntax together with ideas, architectural space design and construction require Geometry together with inspiration and information.

Looking at any architectural database or construction process, indeed, the role of Geometry is clear as a connective-skeleton for any information sets: double check? delete it, and the spatial system "melts", simply disappearing. However, to make it effective in the present era, it needs to be revitalized, together with *Drawing*, which is its major driver, at least in relation to architectural geometry. As a university educator, indeed, I feel a penetrating question mark about *what does drawing mean nowadays in relation to architecture? How (and of course what) can we teach it?*

All in all, something inspiring could arise from looking at the "colors of black" as a metaphor of interdisciplinary, that in our case would mean the power of the network of knowledge based on digital computation: referring again to Calvino, something aiming at supporting and enhancing either our figurative (let us say εικαστικος) or our imaginative (let us say φανταστικος) abilities.

References

1. Barrow, J.D.: Cosmic Imagery: Key Images in The History of Science. WW Norton & Company, New York (2008)
2. Calvino, I.: Six Memos for the Next Millennium. Harvard University Press, Cambridge, Massachusetts (1988)
3. Carr, N.G.: The Shallows: What The Internet is Doing to Our Brains. WW Norton & Company, New York (2010)
4. Cocchiarella, L. (ed.): The Visual Language of Technique. Volume 1-History and Epistemology. Volume 2-Heritage and Expectations in Research. Volume 3-Heritage and Expectations in Education. Springer, Cham (2015)
5. De Kerckhove, D.: L'architettura dell'intelligenza. Testo & Immagine, Torino (2001)
6. De l'Orme, P.: Architecture de Philibert de l'Orme: oeuvre entière … Rouen, Le Palais (1648)
7. Evans, R.: The Projective Cast: Architecture and Its Three Geometries. The MIT Press, Cambridge (1995)
8. Hovestadt, L.: Jenseits des Rasters—Architektur und Informationstechnologie: Anwendungen einer digitalen Architektonik = Beyond the grid—Architecture and information technology. Birkhäuser, Basel-Boston-Berlin (2010)
9. Le Corbusier: Il Modulor + Modulor 2. Mazzotta, Milano (1974)
10. Levy, P.: Qu'est-ce que le virtuel? La Découverte, Paris (1995)
11. Mitchell, W.J.: The Reconfigured Eye: Visual Truth in the Post-Photographic era. The MIT press, Cambridge, Massachusetts (1992)
12. Race, S.: BIM Demystified: An Architect's Guide to Building Information Modelling/Management (BIM). RIBA, London (2013)
13. Rheingold, H.: Smart Mobs: The Next Social Revolution. Basic Books, New York (2002)
14. Ugo, V.: Fondamenti della Rappresentazione Architettonica. Esculapio, Bologna (1994)

Simple Complexities: An Interdisciplinary Approach Towards Computational Design and Architectural Geometry

Marco Hemmerling and Carlo de Falco

> Technical skill is mastery of complexity, while creativity is mastery of simplicity. (Christopher Zeeman)

1 Introduction

The broad application of computational design and construction technologies has widely changed the use and perception of computer software in architecture. Thus, we have seen a shift from mere drawing-tools towards mighty parametric design methods. These tools have allowed many architects to form the conception and design of very complex architectural projects. Hence, many formally intricate buildings have been designed and built over the past years and many of these projects have undergone massive so-called post-rationalization-processes—i.e. methods of (mostly) intelligent geometrical simplification, like the costly triangulation of double-curved surfaces. From today's point of view, such processes, however extremely elaborate in themselves, appear to be a bit anachronistic. They seem like attempts of after-computerization of actually post-modernist design approaches.

However, Patrik Schumacher describes the validity of parametric design as an important movement within the history of architecture when he states: "*Contemporary avant-garde architecture is addressing the demand for an increased level of articulated complexity by means of retooling its methods on the basis of parametric design systems*" [1]. Indeed, much of the attention in the architectural discourse of recent years has been relegated to the field of aesthetics

M. Hemmerling (✉)
Computational Design in Architecture, Faculty of Architecture,
Cologne University of Applied Sciences, Cologne, Germany
e-mail: marco.hemmerling@th-koeln.de

C. de Falco
Politecnico di Milano, Milan, Italy
e-mail: carlo.defalco@polimi.it

M. Hemmerling and L. Cocchiarella (eds.), *Informed Architecture*,
DOI 10.1007/978-3-319-53135-9_3

and high levels of interest in the generation and articulation of complex geometric configurations. Nevertheless, beyond geometry, complexity involves much more invisible layers of interaction as well as interdependencies and non-linear responses. Furthermore, complexity in architecture is often confused with complicated structures. In fact, complexity does not require complicated systems and intricate rules to unfold its potential. On the contrary, complications lead to multiplicative chains of unanticipated effects rather than generating a practical and robust system. In other words, redundancy and overcompensation—as characteristics of a so-called performative system—can solitarily be achieved by focusing on the relational qualities and traceable interdependencies, thus, the conditions of a system.

Neither the discourse on complexity in architecture is new, nor the discussion about the reference system of architectural design. The Post-Modern movement approached complexity by emphasizing the importance of the context for architecture [2] and by contrasting it with the universal Modern science of simplicity and reduction [3]. At about the same time the advent of the trans-disciplinary movements of cybernetics and structuralism introduced the idea of regulatory systems to the field of architecture, which since the 1990s experience a revival, as the available computational means and methods enable a more comprehensive and integral approach towards rule-based design and complexity [4]. Whereas the Structuralism of the 1970s encountered limits in complexity that were insurmountable at the time, today there is much to suggest that the return to this apparently unfinished project is causally connected to information technology, which has opened up new possibilities for dealing with complexity.

Managing complexity, negotiating multiple aspects of a design problem and generating various solutions through computational processes enable the architect today to develop a relational and adaptable architecture of interdependencies. Bernard Tschumi's *statement "Architecture is not about conditions of design, but about the design of conditions"* [5] can be seen as a programmatic testimonial or this reference system of architectural design.

2 Simple Complexities

Architecture has always relied on mathematics [6] to achieve proportioned aesthetics, structural performance, and reasonable construction. Computational tools have now given architects the means to design and build spatial concepts that would have been inconceivable even ten years ago. Against this background the academic project, "Simple Complexities" focused on the early integration of optimization parameters regarding structural performance, physical properties, and material specification as well as aspects of fabrication to inform the architectural design. The curriculum has been conceived and carried out jointly by the Department of Architecture and the Department of Mathematics at the Milan Polytechnic University. The research focused on an in-depth understanding of geometric principles and their mathematical definitions as a starting point for the development of individual architectural projects. Next to the interdisciplinary exchange between Architecture and Mathematics, the intercultural set-up of the design teams (originating from 12 different nationalities) proved to be both challenging and inspiring throughout the process.

The emphasis of the course was put on research-based design-strategies that aimed to unfold hidden complexities of rather simple geometric definitions. 3D-geometries are in first place mathematical objects and as such, they are a sequence of mathematical functions and relations used to describe a set of volumes and the surfaces that constitute their separating boundaries. CAD applications are built using these mathematical concepts but generally, do not unveil them. In order to apply these principles in a comprehensive way, they have to be taught and understood by using the language of Mathematics. Moreover, the constraints that a geometrical shape must satisfy in order to be fabricated conveniently are described as well by mathematical equations. Hence, Mathematics plays a major role for the operability within the design and building process.

Taking these aspects into account, the two-folded teaching approach combined both, the definition and discussion of mathematical principles for the generation of spatial geometry as well as their conceptual, structural, and functional potential from an architectural point of view. The considerations exposed above lead to the awareness of the opportunity for Architects to build a familiarity with the basic concepts of Computational Geometry, in particular with the representation and approximation of curves, surfaces, and volumes, and showed the necessity to build up a firm knowledge of the language appropriate for the discussion of such concepts.

The clear conception of a computation-process, whose rules lead to certain formal and structural consequences, is the necessary first step towards an architecture that is both structurally interesting and systematically coherent. Thus, in this seminar, the students attempted to develop strong and fresh architectural projects from fairly simple methods of parametrical definition of space-defining geometries. And while doing so a conceptual, historical, theoretical, and technical framework was built around the projects to take each design beyond the mere development of geometry and the application of computational tools.

3 Case Studies

The following two projects are chosen as representative case studies for the interdisciplinary approach. The process was organized in a bottom-up strategy, starting with a general research about a self-chosen topic by the students, e.g. folding, projection, developable surfaces, hinge joints, platonic solids, air flow, or minimal surfaces. The projects developed from a basic understanding of the underlying mathematical principles towards the application of a computational design strategy that allowed for the generation of variations within the given geometric typology. Digital and physical models where developed in parallel to investigate and evaluate the properties, aesthetics, and structural performance of the different concepts. Against this background, the teaching concept focused much more on the findings throughout the research process than on the formal results.

3.1 Folding

Rigid-foldable structures are foldable surfaces consisting of rigid panels and hinges, thus can be used for wide variety of deployable structures without relying on flexible materials. The striking elegance of models folded from paper appear not only visually interesting, moreover, they inherit positive structural effects and allow for the creation of multiple pieces from one folded sheet which may save material, fabrication time, and building costs. Hence, the ability to create a structural form from thin, flat materials have made folded structures an ideal candidate for light-weight deployable structures in architecture and engineering [7].

As a starting point of their research, the team (Elena Angeli, Edoardo Corna, Steven Lane) chose the Yoshimura- (or Diamond-) pattern. This origami pattern was named after the Japanese scientist Yoshimura who observed the behavior of thin cylinders folded under axial compression force. He found out that the surface folds of a folded cylinder follow a specific pattern, which is similar to a diamond. The base for this pattern is a deltoid that is folded along a diagonal. Deltoid edges are folded as a mountain fold, while one diagonal is a valley fold. Because the Yoshimura-pattern is obtained through the in-extensional deformation process, it can be considered as the candidate of a foldable cylindrical shell [8]. It must be noted that while the buckled pattern is developable, it cannot transition to the unstrained cylinder without undergoing extensional deformation and further collapsing would require vertices to move through the material. Based on these findings, a computational tool was developed that allowed for the generation of cylindrical shell structures, applying the Yoshimura-pattern.

The concept of this cylindrical shell structure is often referred to as Schwarz lantern after the German mathematician Karl Hermann Amandus Schwarz (1880). Figure 2 describes the mathematical definition and geometric features of this particular origami shape (Fig. 1).

The theory was transferred to a computational definition in Rhinoceros/Grasshopper[1] and then applied to a series of physical models, fabricated from paper and cardboard to prove the concept and investigate the structural performance of the cylindrical shell structure: The higher the resolution (number of folds), the more rigid and stable the structure performs. The folding principle was finally applied to an ACM board (Aluminum Composite Materials) in order to scale the principle and related the concept to possible architectural components (Fig. 3).

[1]www.rhino3d.com, www.grasshopper3d.com.

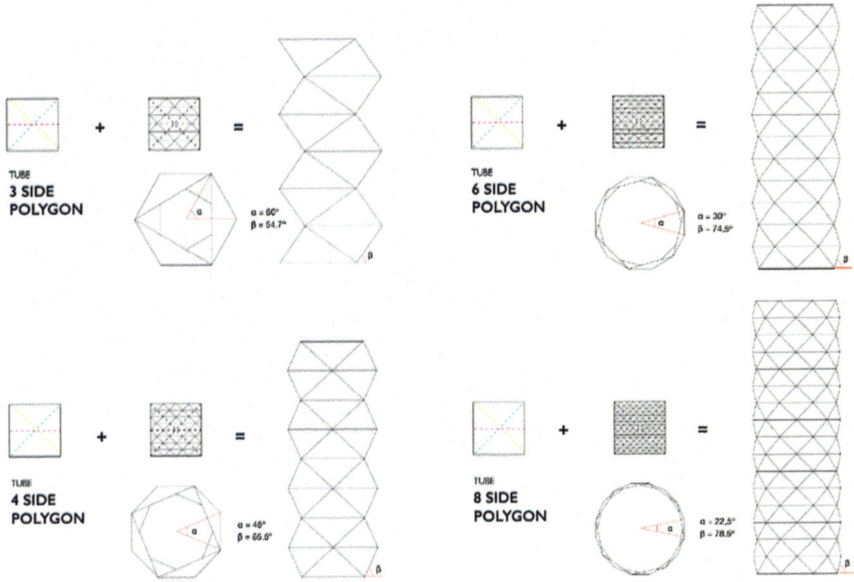

Fig. 1 Geometric definition of cylindrical shell structures with different resolutions of the Yoshimura pattern

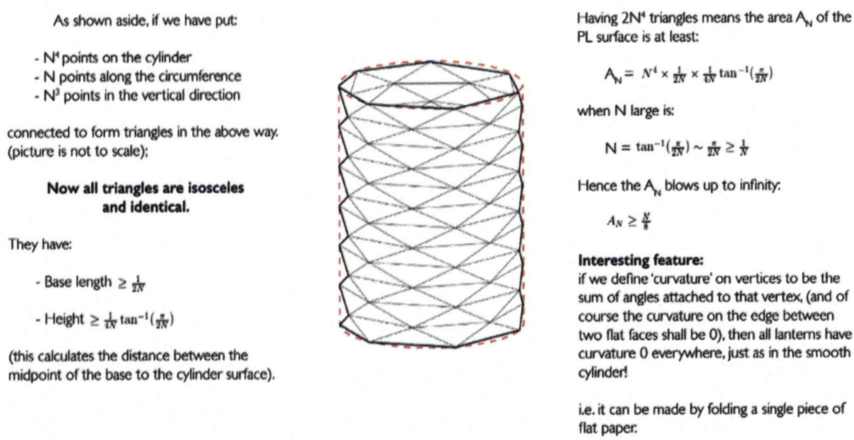

Fig. 2 Geometric definition of the Schwarz lantern

Fig. 3 Physical prototypes in cardboard (*left*) and folded ACM-panel (*right*)

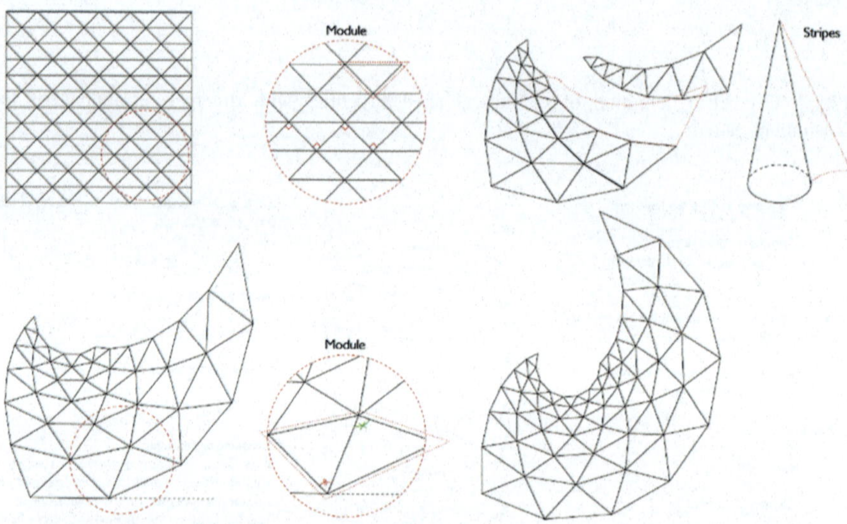

Fig. 4 Yoshimura pattern applied to a conical geometry

In a final development step the folding principle was applied also to conical shapes, taking into account the deformation/scaling of the Yoshimura-pattern while maintaining the previously defined properties (e.g. unfolding the pattern to a single sheet) (Figs. 4, and 5).

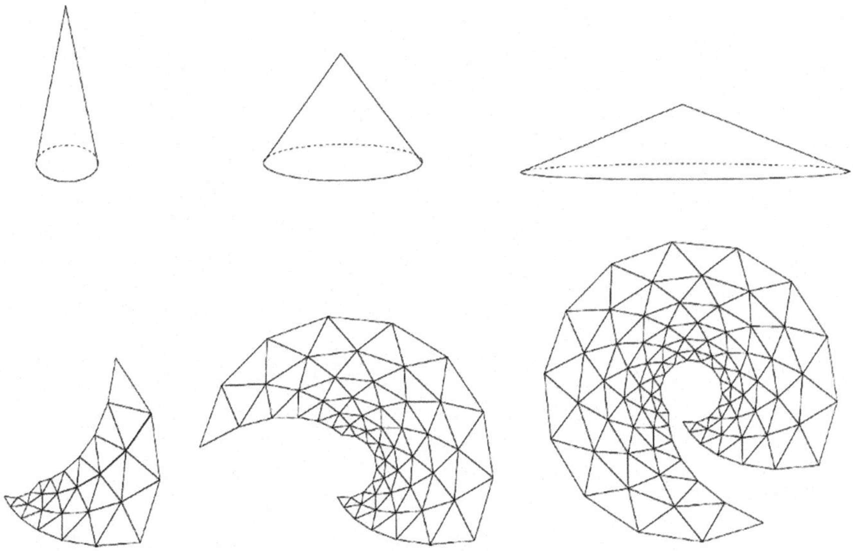

Fig. 5 Variations of the conical geometry

3.2 Projection

The concept of projection is well known and applied in architecture to represent and communicate design concepts. Architectural drawings are made according to a set of conventions which include particular views. For most of its history, architectural drawing has relied heavily on techniques of graphical projection, understood as protocols by which images of three-dimensional objects are projected onto a planar surface. Some form of graphical projection—typically parallel projection, orthographic projection, or perspective—are employed on the vast majority of architectural drawings, either drawn by hand or automatically generated by computational tools.

In orthographic projection, the projection rays are parallel to one another and perpendicular to both the image plane and a dominant plane of the object depicted. This type of projection results in two-dimensional representations of the projected object. In axonometric projections, the projection rays are parallel to one another and perpendicular to the image plane—but in no specific relationship to any dominant plane of the object depicted. Hence, the resulting image appears as a three-dimensional representation. In contrast to these two parallel projection types, the perspective is an approximate representation on a flat surface of an image as it is perceived from a particular fixed viewpoint. As such, the projection rays converge at a "station point" representing the disembodied eye of a viewer.

The notion of projection was introduced in Euclidean geometry to denote the projection of the Euclidean space of three dimensions onto a plane in it, like in the phenomenon of the shadows cast by real world objects on the ground or the image of a geometric figure reproduced on a line, plane, or surface. More abstractly, a projection is a mapping of a set into a subset.

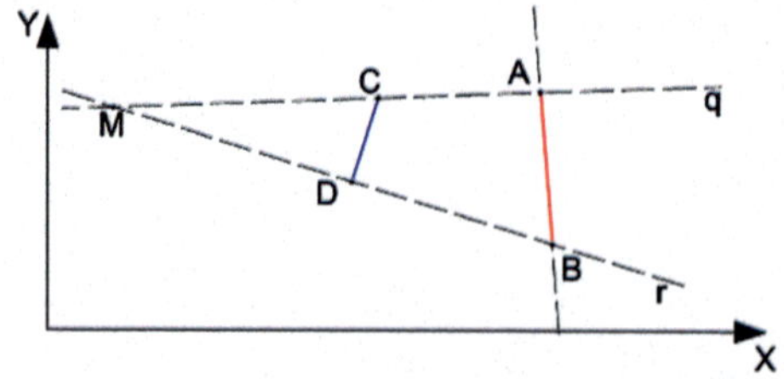

1. Finding a line that goes through M and A (q)
2. Finding a section of lines r and q (point C)
3. Providing the same procedure for point D
4. Final formulas :

$$x_C = \frac{y_m - \frac{y_a - y_m}{x_a - x_m} x_m - n_r}{k_r - \frac{y_a - y_m}{x_a - x_m}} \qquad x_D = \frac{y_m - \frac{y_b - y_m}{x_b - x_m} x_m - n_r}{k_r - \frac{y_b - y_m}{x_b - x_m}}$$

$$y_C = k_r \left[\frac{y_m - \frac{y_a - y_m}{x_a - x_m} x_m - n_r}{k_r - \frac{y_a - y_m}{x_a - x_m}} \right] + n_r \qquad y_D = k_r \left[\frac{y_m - \frac{y_b - y_m}{x_b - x_m} x_m - n_r}{k_r - \frac{y_b - y_m}{x_b - x_m}} \right] + n_r$$

Fig. 6 Projection in analytical geometry (2D)

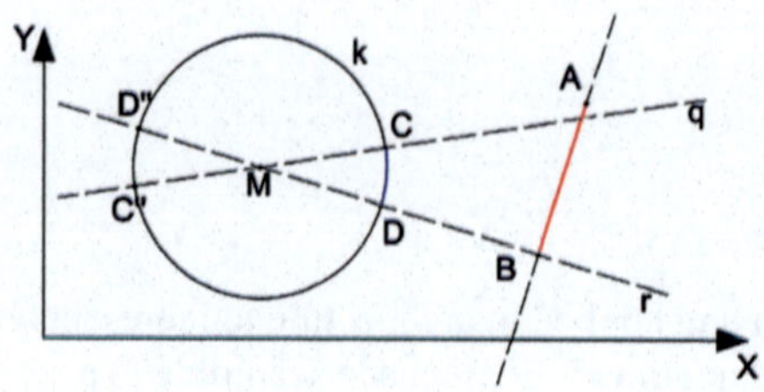

1. Finding a line that goes through M and A (q)
2. Finding a section point of line q and circle k (C)
3. Replacing Yc in equation of circle
4. Final formulas (2 solutions - circle and line can have 2 section points) :

$$x_c = -x_m + \frac{r}{1 - \frac{y_a - y_m}{x_a - x_m}} \qquad y_c = \frac{y_a - y_m}{x_a - x_m} \left[-x_m + \frac{r}{1 - \frac{y_a - y_m}{x_a - x_m}} \right] + y_m - \frac{y_a - y_m}{x_a - x_m} x_m$$

Fig. 7 Reverse projection on a circle (2D)

Against this background, the group (Milan Dragoljevic, Federico lannarone, Martina Jamroz, Alfiia Khabibullina) studied various types of projections to understand the geometrical dependencies and mathematical definitions in order to develop a computational tool that allows for the design of a light object (e.g. projector, lamp) based on these principles (Figs. 6, and 7).

During their research, the students came across the concept of anamorphosis[2] which deals with a distorted perspective. The concept is based is on the process of

[2]The word "anamorphosis" is derived from the Greek prefix *ana-*, meaning back or again, and the word *morphe*, meaning shape or form.

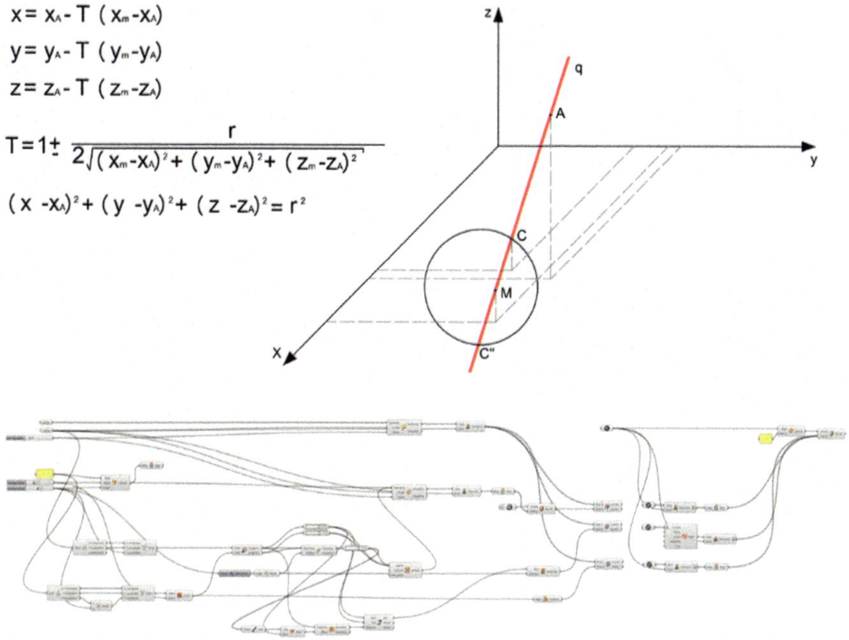

Fig. 8 Mathematical (top *left*), geometrical (top *right*) and parametrical definition (*below*) of the design tool

Fig. 9 Projection of a defined shadow pattern from the spherical lamp to an orthogonal space (X-, Y-, Z-planes). In this case the position of the viewer is not important, because the projected pattern is coplanar to the given planes

Fig. 10 Introduction of an object, e.g. a cylinder in the scene. The principles of projections are applied two times in this case. Firstly, from the lamp to the given planes of the space, and secondly from the environment to the viewer. In this case the position of the viewer is precisely defined in one spot. Hence, the projected image is not coplanar with the environment anymore, because of the presence of the cylinder in the scene

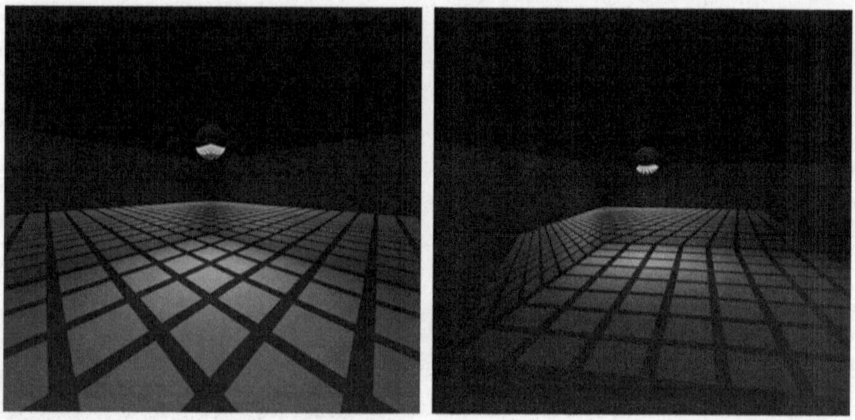

Fig. 11 Projection of patterns belonging to a plane which is not aligned with the given planes. Also in this case the position of the viewer must be fixed to experience the pattern as intended

greatly distorting an image only to have it revealed either from a single vantage point (oblique perspective)[3] or from its reflection on a mirrored surface (catoptric).

By translating the findings from the initial mathematical definition of a reverse projection on a circle (2D) to a sphere (3D)—combined with the concept of the

[3]Leonardo's Eye (Leonardo da Vinci, c. 1485) is the earliest known definitive example of perspective anamorphosis in modern times.

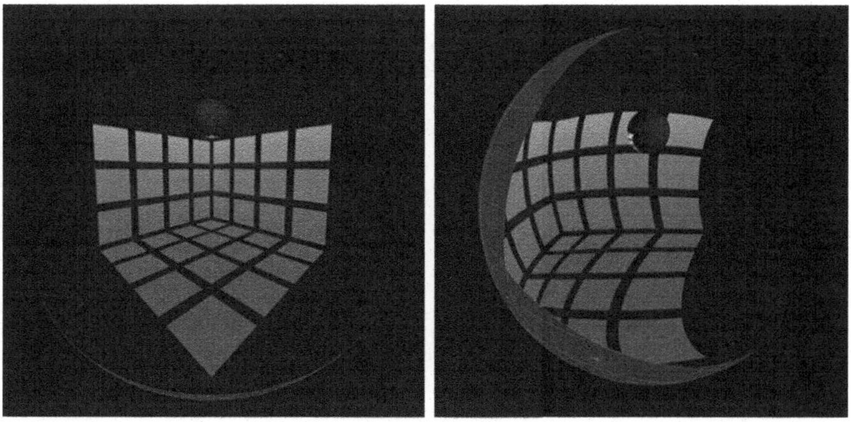

Fig. 12 Projection of rectangular patterns onto regular smooth surfaces, such as a spherical environment. The position of the viewer is fixed

Fig. 13 Projection of rectangular patterns onto irregular (free-form) surfaces. The position of the viewer is fixed

oblique perspective anamorphosis—a parametric model was set-up in Rhinoceros/Grasshopper (Fig. 8).

As a result, the computational design tool allowed for the generation of various projection scenarios, including: (Figs. 9, 10, 11, 12, and 13)

For the virtual tests of the different projection scenarios the 3D-models were exported to 3DS-Max[4] to simulate the light ambiance, the shadow mapping, including the definition of the light source, the shape and pattern of the lamp as well as the geometry of the environment. In the final step of this research project, a physical prototype of the lamp had to be developed in order to verify the theoretical

[4]http://www.autodesk.com/products/3ds-max.

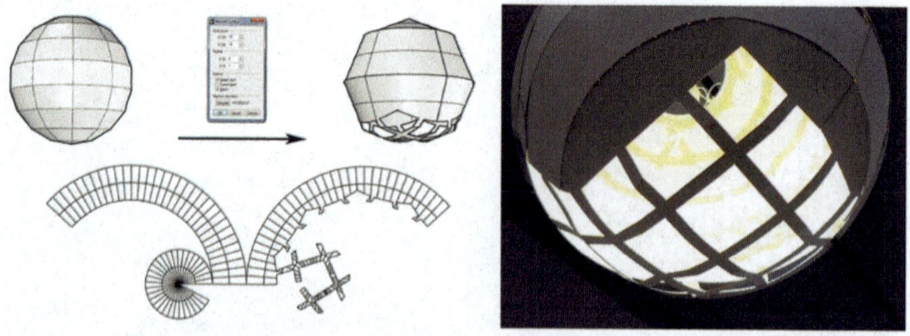

Fig. 14 Approximation of the geometry and unrolled-cutting pattern (*left*) for the production of the physical prototype (*right*)

concept. For the production of the spherical lamp prototype, three strategies were taking into account:

1. Unrolling the geometry to a cutting pattern, which will be laser-cut from paper and then folded to the 3D-shape. Hence, the double-curved geometry has to be approximated by developable surfaces.
2. Slicing the model in horizontal layers for the production of cardboard sections (also laser-cut), which will then be glued on top of each other.
3. Offset of the surface-model to generate a watertight volume-module that allows for 3D-printing (STL-file).

The definition of the fabrication files for all three production methods could be included in the Grasshopper logic and then be generated automatically. The first two approaches, however, results only in an approximation of the original, double-curved geometry as they will be produced from flat sheets of paper/cardboard. The third option (3D-Print) would for sure represent the original geometry in the best way and could include the detailed fixture for the light source quite nicely but it is by far the most expensive option in the desired scale of the prototype (lamp-diameter = 50 cm). At the end, the approximation of the sphere by segments of developable stripes (option 1) was chosen for the production of the first prototype, as it is both inexpensive and material-efficient. The physical set-up of the projection finally proved the computational design strategy (Fig. 14).

4 Conclusion and Outlook

Managing complexity is for sure one of the major tasks we are facing in architectural research and practice nowadays. A design approach which tries to implement and connect the available and relevant information in an all-embracing process is obviously very complex. Therefore, the Master course connected complexity deliberately with a research-based bottom-up design methodology to constitute an attainable learning base for the students. As a result, almost all of the projects reached a high level of complexity that started from a fairly simple analysis of a phenomenon or mathematical problem. In order to generate an accessibility to the chosen topic, the abstraction of the principle was an important first step (and a threshold for most of the teams as it appeared to be quite different from a standard design process). The definition of a mathematical model that represented the basic idea served as an essential starting point. Hence, the more profound the model was elaborated, the more potential it inherited for the further design process. Furthermore, the approach allowed for the successive development of a resilient system that is able to generate variations (as a base for design decision making) and to increase the level of complexity throughout the design process gradually.

Yet, the case studies also showed that the integration of parameters and requirements of functionality, efficiency, and aesthetics as well as structural performance shifts the focus from a purely formal design practice to an optimization process that relies on a systematically coherent approach. The understanding of the mathematical definition laid the solid foundation for the generation of a computational design concept and allowed for an accessible handling of complexity throughout the whole design to build process. Likewise, the development of physical prototypes served as a relevant source of information for the development of robust design solutions that take into account aspects of materialization, fabrication, construction, and assembly.

The findings shall be integrated in future projects, both as a design and teaching method that can be applied to new topics. Moreover, the projects shall be developed further towards a 1:1 realization in architecture to prove particular computational design and construction strategies.

Acknowledgements The interdisciplinary Master course "Simple Complexities" was conducted at the Politecnico di Milano in 2015/16 by Prof. Marco Hemmerling and Prof. Carlo de Falco with the support from Alessio Mazzucchi and Jacopo Corno.

Participating students: Zhasmina Aleksandrova, Sofia Ambrosini, Elena Angeli, Matteo Assecondi, Alberto Genolini, Margalida Bibiloni, Chiara Bona, Antonella Branizza, Manuel Catrileo Herrera Ceccon, Edoardo Corna, Nikolet Dishkova, Milan Dragoljevic, Sheetal Gandhi Mugdha, Tsovoo Gansukh, Radoslav Georgiev, Roberto Gomez Muller, Karol Gwiazdowski, Federico Iannarone, Martina Jamroz, Alfiia Khabibullina, Kamila Khalitova, Anastasiia Khech, Steven Lane, Julio Lara Monjaraz, Domenica, Lecca, Gaia Marfoni, Davide Masserini, Nalisha Men, Avinash Arvind Mewada, Anna Mikhailova, Golnaz Nouri, Andrea Palermo, Poll Seguí, Daniel Polo, Guillermo Molina, Yehan Wang, Ivan Yotsov.

References

1. Schumacher, P.: The Autopoiesis of Architecture: A New Framework for Architecture, vol. 1, Wiley & Sons, London (2010)
2. Venturi, R.: Complexity and Contradiction in Architecture. Museum of Modern Art, New York (1966)

3. Jencks, C.: Nonlinear Architecture. New Science = New Architecture? Architectural Design 129 (1997)
4. Valena, T.: Structuralism Reloaded. Rule-Based Design in Architecture and Urbanism. Edition Axel Menges, Germany (2011)
5. Tschumi, B.: *Responding to the question of complexity*, In: Andrew Benjamin (ed.) Complexity. Art, Architecture, Philosophy, Journal of Philosophy and the Visual Art, Nr. 6. London (1995)
6. Pottmann, H., Asperl, A., Hofer, M., and Kilian, A.: Architectural Geometry. Bentley Institute Press, Exton (2007)
7. Jackson, Paul: Folding Techniques for Designers: From Sheet to Form. Mac Win Pa edition, Laurence King Pub (2011)
8. Hunt, G., Airo I.: Twist buckling and the foldable cylinder: An exercise in origami. Int J Non-Linear Mech 40 (2005)

The Algorithms-Aided Design (AAD)

Arturo Tedeschi and Davide Lombardi

1 Input/Output Data

To face the unprecedented complexity of the real world, designers must get a deep control and understanding of **datasets** and, most of all, they have to find new strategies to collect data and process them in order to inform the design that is to be formed. From running-shoes to high-rise buildings or bridges, data is crucial to develop ambitious projects, which crucially creates articulated entities, not as a representation of complexity but as the solution for complexity.

The current (and still evolving) stage of digitalization applied to architecture demonstrates to us that digital tools are useful to explore a potentially unlimited number of design solutions in order to find the best solution to a specific problem. The design project does not find its essence in a priori defined specific wills but it is the outcome of a process in which an important role is played by new forces that was impossible to describe and control so far. The inspiration and the guidelines of the design intent are no longer collected from the rules of an architectonic tendency or from artistic influences but their reasons are now found in sets of data gathered from the environment. We not only refer to a 1-to-1 scale built environment but primarily to the realm in which physical forces, structural strain and stresses, or the molecular properties of materials exist and carry out their functions. Furthermore,

---

Algorithmic design is not simply the use of computers to design architecture and objects. Algorithms allow designers to overcome the limitations of traditional CAD software and 3D modelers, reaching a level of complexity and control which is beyond the human manual ability. (Arturo Tedeschi).

A. Tedeschi (✉)
Computational Designer, Milan, Italy
e-mail: info@arturotedeschi.com

D. Lombardi
Department of Architecture, Xi'an Jiaotong-Liverpool University, Suzhou, China

M. Hemmerling and L. Cocchiarella (eds.), *Informed Architecture*,
DOI 10.1007/978-3-319-53135-9_4

the environment is also the space in which huge masses of agents are spread out according to flock-based behaviour and responding to external stimulations defining new paths through cities and places. This enlarged environment has a strong multi-scalar future, from the microscopic to the urban scale.

We can recognize the birth of a new kind of profession—the **computational designer**—that deals with a wide range of computational tools that allows him to control and investigate the defined new environment. Data represents the real innovation introduced by the digital turn and embody the passing of representation as the main tool to design and show the project idea. Due to their neutrality and to the inner mathematical nature, the use of data within the computational setting has to be driven by mathematical rules. The data represents the **input** and the **output** of the design process; the start and the end point of an ideal path made by consecutive logic steps defining an algorithm. This is the area in which the **Algorithm-Aided Design (AAD)** is born. It explores new fields gaining its power from the ability to describe the complexity of the real world through numbers and mathematical functions. The AAD is funded on the analysis of the factors that affect the project itself and, when translated into data, it analyses and uses them in order to inform the process and to optimize the outcome according to a determined fitness function. The AAD has the strength to contain information that would be impossible to control and use through the classical graphic representation.

2 Every Step Is Data: New Balance 3D Printed Shoes

Among the number of possibilities provided by this approach there is the almost endless ability of project customization. Both in architecture and in product design or fashion design, it is possible to work with very detailed input data. We can mention here the running shoes research by New Balance, where the interaction among different subjects as fashion, digital fabrication, material behaviour and Algorithm-Aided Design provide an innovative outcome in terms of sport performances. The **input** data inform the design which embeds the **output** data controlling the fabrication process.

The midsole shape, manufactured by 3D printers using an innovative elastomeric powder, is obtained by recording the pressure applied by the runner through his running session so that *"every step, every impact is data"*.

The collected data are used to create density differentiation across the midsole to support the runner in those points that are more stressed. The pseudo-chaotic final result hides a real scientific approach completely customized based on the specific needs of the runner.

The traditional drawing as a tool to investigate possible design solutions appear insufficient to elaborate both the huge quantity and the kind of information that intervene to modify and to describe the final shape. Pencil and paper are not able to describe, inside a static draft framed in a standard sheet, the innate dynamicity of forces and agents that represent the principal inputs in contemporary design.

Further, always more often the idea of representation is substituted by the idea of simulation. The use of a number of different parameters requires to simulate the results obtained through the definition of sets of them and to evaluate these results in order to be discarded or approved to the next algorithmic stage. This workflow states again the importance and the superiority of the whole process if compared with single solutions and parameters.

3 Topology Optimization: 3D Printed Steel Bridge

A forthcoming project, the 3D printed steel bridge, is already proving in the design stage the potentials of the AAD applied to large scale projects. A company called MX3D, specialized in using robotics to 3D print, and Dutch designer Joris Laaramhave designed a metal structure, which will span a canal in Amsterdam, that will be printed in situ by robotic arms.

Thanks to support and software from Autodesk, the systems are able to weld metal or print resin, in mid-air, without support structures or size constraints.

Aided by the geometry of the overall bridge design (based on a topology optimization process) the robots can print the metal horizontally without bending and falling due to gravity. Tim Geurtjens of MX3D says: Autodesk is working on a new topology optimization process which goes further than most traditional optimization software. It will take into account more parameters and constraints. If you incorporate those parameters into the design software, the whole process will be more efficient and it makes the bridge stronger".

4 Algorithms as Creative Tool: The Cloudbridge

Merging computational techniques with a natural architectural language, the Cloudbridge by A > T (with the collaboration of Marcin Kasiak) creates a non-linear path suspended between two points. A specific algorithm was used to design a cloud-shape grid that balances the asymmetric loads of the footbridge. The overlapping of transparent geometric forms joins the bridge together, offering an experience that helps to promotes a journey rather than just a passing from 'a' to 'b'.

The Cloudbridge is an attempt to imagine a new way to design a bridge. Arturo Tedeschi says: "Nowadays engineering and architecture are evolving just by improving, in small steps, the "state of art" and designers are gradually losing their visionary attitude". This is also evident in technology, automotive and product design. Cloudbridge has the ambition to be a visionary and playful manifesto which invites to explore new trajectories in architecture, rethinking the way we conceive buildings. The project also suggests a kind of architecture which is not only integrated to nature, but is a part of natural landscape. The project is inspired by the radical approach of 60's avant-garde (in particular by the Italian team Super studio), by natural suggestions, as well as the contemporary iconic architecture (Figs. 1, 2, and 3).

Fig. 1 The Cloudbridge (Design by A > T)

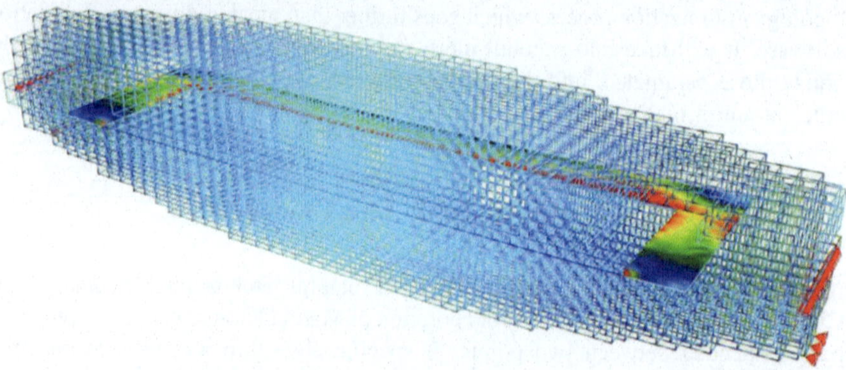

Fig. 2 The Cloudbridge. Structural analysis (Design by A > T)

Roughly speaking, the algorithmic modeling allows designers to create complex shapes by the computer without clicking the mouse, without any manipulation of digital forms. The designer creates a diagram (by a specific software) which embeds a set of procedures and instructions that "generate" the form: the shape is not actually "modeled", but "emerges" from a series of instructions.

An algorithmic-software could be imagined as a blender and an algorithm as a recipe. "We put the ingredients into the blender with specific quantities and combinations, then we push the start button and our "shake" emerges. In the Cloudbridge bridge the ingredients were dimensions, weights, construction

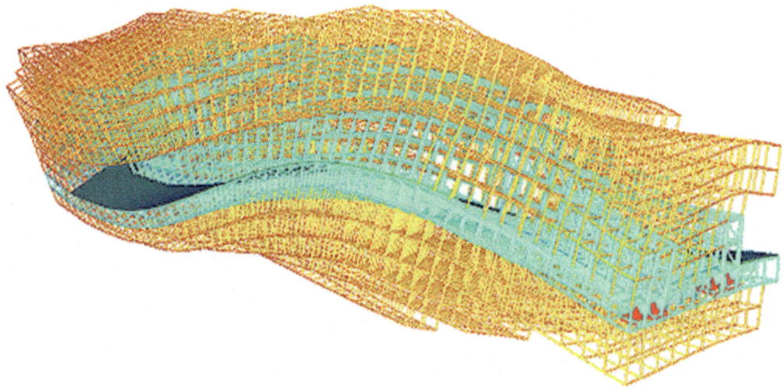

4th Vertical mode of vibration at 5.51 Hertz.

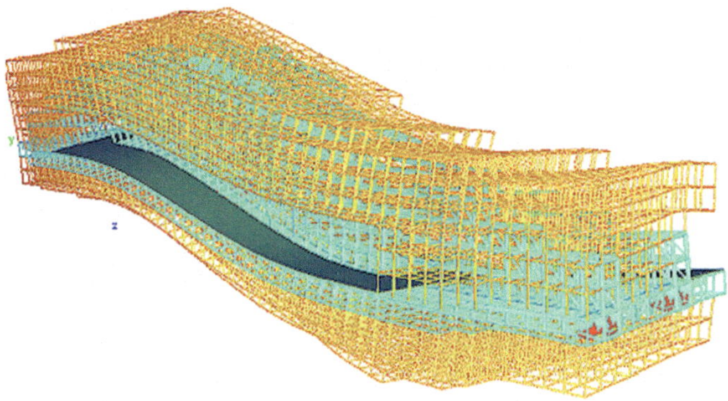

5th Vertical mode of vibration at 6.20 Hertz.

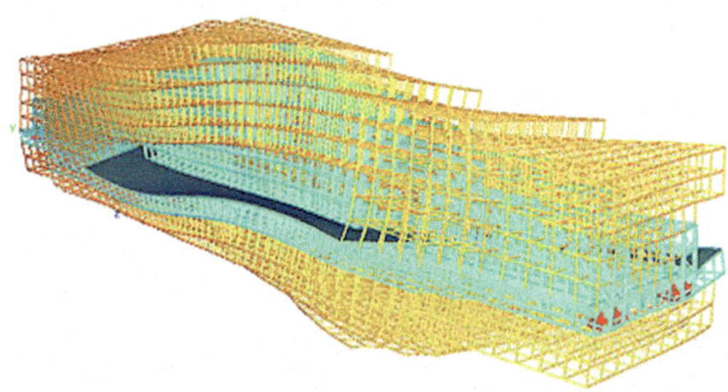

6th Vertical mode of vibration at 6.83 Hertz.

Fig. 3 The Cloudbridge. Vibration analysis (Design by A > T)

constraints, shape, and path's length. The software computes all the ingredients according to the algorithm and the final output is an optimized 3D-grid that bears and balances the asymmetric loads of the path. It means that, for every arbitrary path, the algorithmic software generates the best structural grid. Such a strategy enabled us to avoid the imposition of prefixed forms and suggests a performance-driven design.

The AAD approach can lead designers to new territories they must not be afraid about. Nevertheless, such approach should avoid the dangerous seduction of the so called "automatic design" that produced, so far, just mechanisms rather than architecture. Fortunately, algorithms are not blind processes and designers can balance the "ingredients" in order to give their personal contribution. Architecture - especially regarding its relationship with natural environment and with the realm of perception - should deal with ancestral, symbolic contents that can be basically "solved" by the human intelligence.

Building Information Modeling: The 'C' in BIM

Boris Bähre

1 Introduction

When looking at worldwide building processes today, a common and concerning pattern emerges. A large number of projects are not finished on time, exceeding the estimated costs by significant amounts, and/or are not completed with the expected quality. Even if building professionals start projects with the best of intentions, using newest technology or state of the art management, they often appear to lose grip and oversight on the building process along the way. While disturbing and stressful yet still manageable in smaller projects, these issues can cause enormous problems in large scale projects. Approximately one of four mega projects (projects with building costs of 1 billion U$ or more) are facing significant delays and cost increases. Recent examples include the Seattle tunnel project in Seattle, Canada (time delay of about 4 years and cost overrun > 200 million US$, still ongoing), the opera house 'Elbphilharmonie' in Hamburg, Germany (time delay of 7 years and a cost overrun about 750 million US$, just finished) and the new Berlin International Airport BER in Germany (time delay of about 6 years and a cost overrun of about 3.6 billion U$, still ongoing). It seems that there is an increasing tendency of authorities and the public to accept a 'first build and then fix errors' culture as unavoidable [1].

Zooming in on the reasons for delays and cost overruns, building professionals already realized that most time and money influencing questions are raised during or even after construction of a building ends. Instead of limiting design and planning to

B. Bähre (✉)
Architecture, BIM, Computational Design @ BE#,
Groningen, The Netherlands
e-mail: mail@bbaehre.com

M. Hemmerling and L. Cocchiarella (eds.), *Informed Architecture*,
DOI 10.1007/978-3-319-53135-9_5

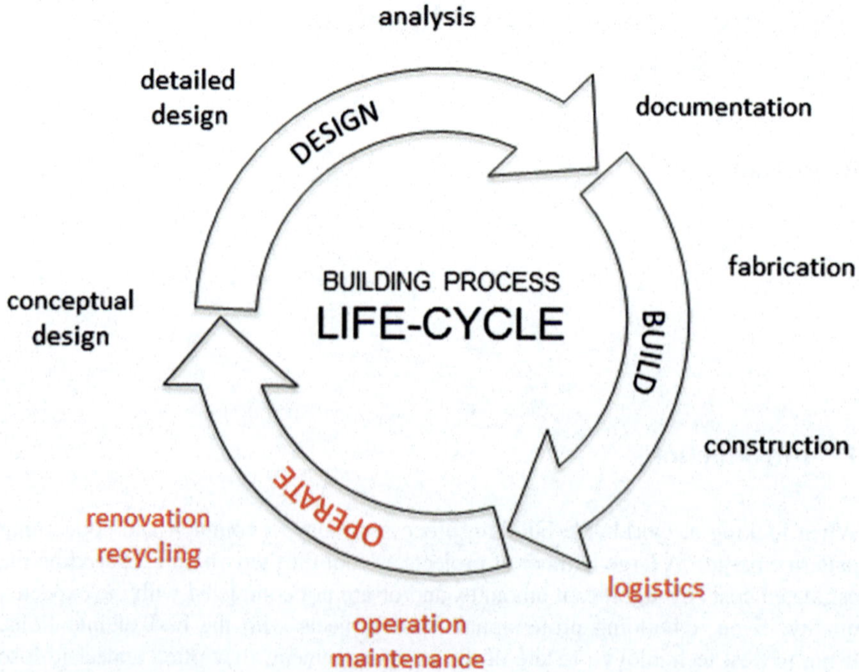

Fig. 1 The 'life-cycle' of building process steps

the phases in between conceptual design and construction (see **black** colored items in Fig. 1) we need to look into the whole life cycle integrating logistics, operation/maintenance, and renovation/recycling as well (see **red** colored items in Fig. 1).

The late building process phases are the moments when we waste the most time and money by changing a 'not thought through' design on site. Introduced by a picture brought up by HOK's chairman and CEO Patrick MacLeamy in 2012 [2] building industries started to realize that by investing 1US$ in planning they spend 20US$ in construction and almost 60US$ in operations and maintenance. This means that the most effective way to save money and effort in the building process in its whole is to rethink and optimize the early stage planning. Every dollar saved by better and more thought through planning will be saved by factor 60 in operations. It is important to understand that the mentioned factors are inheriting changing and waste costs of material and efforts used to correct planning mistakes as well—a double reason to optimize the planning phases in the beginning.

The idea of keeping late phases in mind while starting to design is not new. There were numerous efforts in trying to introduce integration into design and building processes. The efforts more or less failed as the market did not really ask for it and because the technology was not able to handle the amount of needed data in a user friendly enough way to be useful.

2 The Introduction of BIM

Ignited by the Lehman Brother bank collapse and the following housing market crisis around the year 2008, the building market started to ask for more efficiency. Supported by new digital building process planning tools, the think-tank Silicon Valley brought most of the existing ideas and developments related to building processes together and introduced a new method of processing to the market.

The term BIM (Building Information Modeling) was born. There were and still are almost as many definitions about what BIM really is. In this article, I will focus on BIM defined as a method defined as:

- BIM seen as a method describes a collaborative process of designing a building by taking all relevant 'life-cycle' (Fig.1) data into account.
- All information is stored, managed and reusable through a digital 3D model.
- All information is continuously available and all participants have access to the same data.

Already widely introduced in aviation, automobile, and boat industries worldwide, we still do not see a worldwide introduction of BIM in building industries. Although the introduction has been strongly supported by some countries, it was not at all supported by others. What could be the reason within the building world to hesitate with the move into integral building processing? To search for an answer it is helpful to look at key elements of a method like BIM (Fig. 2):

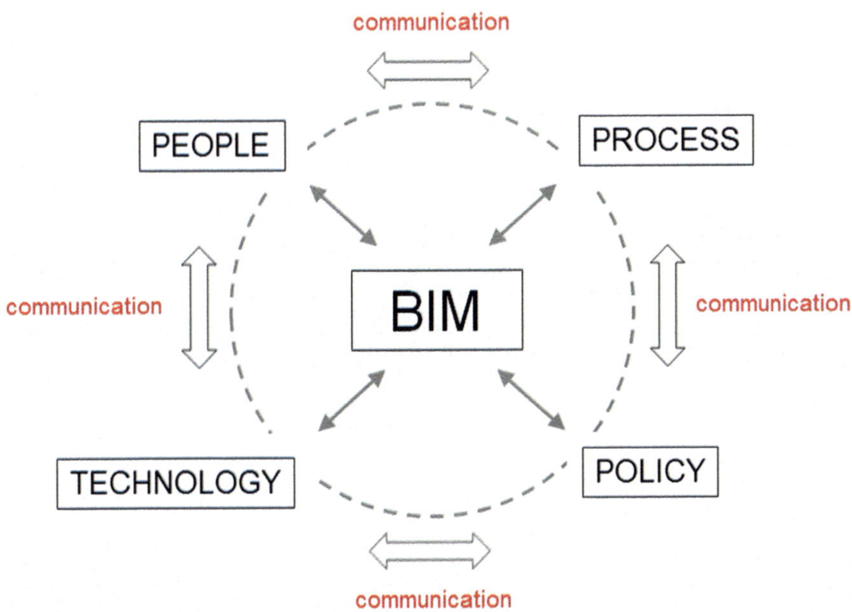

Fig. 2 BIM as a method based on Process, Policy, People and Technology—held together by communication

BIM seen as an integrated method is built on four major legs: Process, Policy, Technology, and People. Using BIM means to consider and develop all four 'legs' equally and to use suitable continuous communication to keep the process running. Leaving just one leg underdeveloped or stopping the flow of communication in between will significantly distort or (in worst-case) kill the whole process. This means that building partners need to communicate much more and quite differently throughout the building process. Yet, building partners often lack the skills to do so. In my view, this is one of the major reasons that the building world is unwilling and/or unable to adopt the BIM method.

In the following I will give an overview of the major elements of an integrated building design process and will show what important role communication (the 'C aspect') will have to play if we are aiming for successful integration in methods like BIM.

3 Leg 1: Process

3.1 Process Organization

Traditional planning methods have a tendency to lose important information during the process steps as it is difficult to check all incoming and 'stacked' data due to short revision times. An integrated approach tries to keep all data as long as possible in a common digital virtual model, where ALL participants are at same eye level able to access the same data 24/7 if needed (Fig. 3).

The '**C**'-Aspect:

Paper printouts while using a collaborative model are just snapshots of an active project and need to be understood as disconnected data. The moment you print, someone else may already have added new information to the active model which is not shown on your printouts. It is highly recommended to minimize printing on paper as much as possible.

Optimizing the time-consuming verification process will be the future goal. Most probably it will be the owner asking all participants including authorities and companies to adapt new and already existing digital tools like game engines, virtual reality, and augmented reality to walk through the building step by step while digital construction happens.

3.2 Frontloading

To avoid costly and timely changes on the already started building site, all planning activities need to take place much earlier than in the traditional building process. All

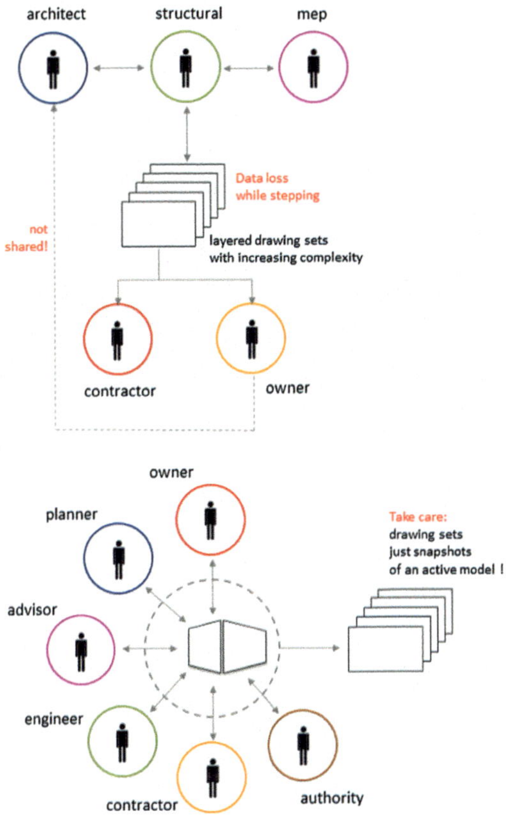

Fig. 3 Traditional (*above*) vs. integrated (*below*) process organization

participants need to sit around 'one table' discussing and adjusting together from day one. The required amount of earlier, concentrated effort, hours and participation has to be set by the owner and accepted by everyone participating. It is important to mention that the overall planning costs will need to stay the same. The only difference is that they need to be distributed differently such that they are concentrated on earlier moments in the planning process (Fig. 4).

The '**C**'-Aspect:

In parallel processing, far more people and decision making in the same time are needed. The increased density of information and earlier decisions to be made will ask for better and different communication skills. The ability to manage a large number of people organized in teams participating in the parallel process will be needed to get the most out of this kind of processing.

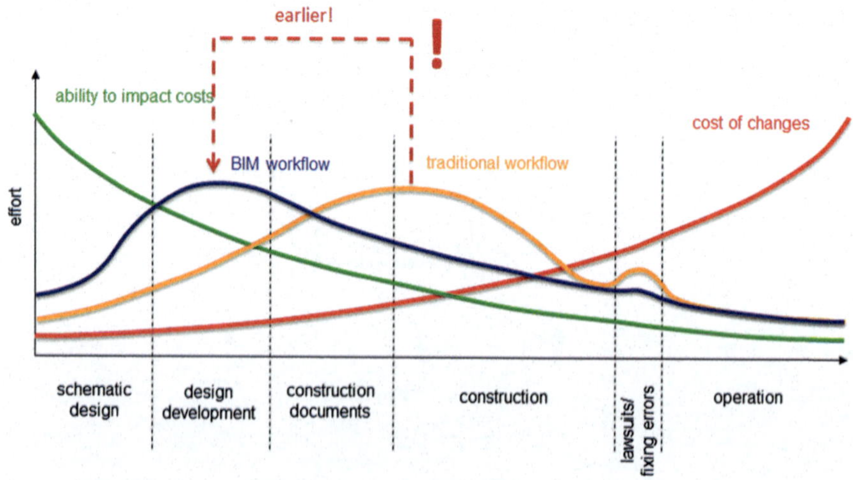

Fig. 4 Effort in traditional and BIM workflows related to costs (*Modified on Ref.* [3])

3.3 Linear Vs. Parallel

Compared to a traditional linear process, BIM process steps are to be organized in parallel steps. Verifying and changing (see the red-dotted line in Fig. 5) will be possible far earlier and therefore with less negative time and cost impact. Placing the bidding phase at the earliest possible moment is crucial for short verification times, the ability to change the model in an early stage, and the possibility to complete the project earlier.

The '**C**'-Aspect:

There are tendencies to avoid traditional bidding at all. Looking at the alternative contract forms (see Policy/Contracting) it can be an alternative to include a bidding phase into a 'Fix Price contract' tender before the project starts. The trust and commitment needed by the owner will ask for an agreement towards a long-term relationship based on the results that the team and contractors will deliver in the end.

4 Leg 2: Policy

4.1 BIM Execution Manuals

Overview of the contents within a BIM execution manual:

- common project goals
- used technology during the phases

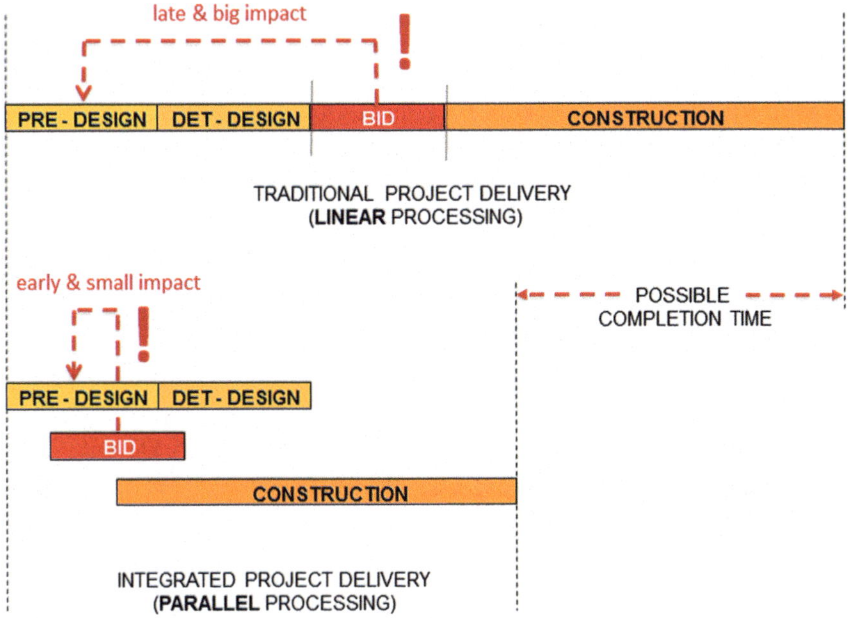

Fig. 5 Linear vs. parallel process flow

- roles and responsibilities
- working together strategy
- data management
- copyright and security
- software
- quality management

An integral approach without clear and a priori defined regulations and commitments between the participants will most likely fail. Countries using BIM introduced nation-wide BIM execution standards. Set up from templates, adjusted to the needs of the project and signed by everyone before the project starts, they are a needed and reliable base for contracting, securing, and working together. Worldwide execution plans are mostly based on the American 'Penn State BIM Execution planning guide' or the British 'AEC UK BIM Protocol project execution plan'. Both templates are downloadable via online sources for free.

The '**C**'-Aspect:

Traditional project handbooks or similar CAD standards are mostly used in larger companies or projects and often just between the most important planning partners. Using BIM needs far better preparation and agreement from all participants. Developing a company's first BIM handbook requires a significant amount of preparation time for setting up and testing before being able to use the standards in a real project.

Due to complexity and changing participants, each building project needs its own BIM Handbook, adjusted from an office wide template. There need to be an agreement whose handbook is used for the whole team and it must be mandatory that every partner signs the document before starting to participate. It is often underestimated how much trust and agreement are needed upfront.

4.2 Contracting

Overview about alternative contract types and forms:

- **DBB**: Design Bid Build (the traditional contract form, widely known)
- **DB**: Design & Build (already known contractors take over the major project responsibility in design and construction)
- **DB CMR**: Construction manager at risk
- **DB FP**: Fixed price contract form
- **DB CPF/GMP**: Cost plus Fee/Gar. max Price

Starting at traditional Design Bid Build (DBB) contract forms the market already developed Design Build (DB) contract types where contractors take over the major project responsibility in design and construction once construction begins. As the design of projects should be finished before construction starts DB contract forms became the most used ones when choosing for BIM. New DB forms such as CMR, FP, CPF/GMP are promising variations of contracting. Even if not landed in all countries so far, they represent interesting options we most likely will see more often in the future.

The '**C**'-Aspect:

Most of the BIM suitable contract forms are based on contracting on a different understanding regarding trust and reliability. It is the owner's aim to get a high quality project within time and cost estimates and most of all with extra effort kept as minimal as possible. Planners and contractors need to understand that it is no advantage to acquire a BIM based project by not telling the truth upfront. As the BIM process is discovering faking or cheating in early stages of planning, there is enough time to change partners without too much cost. Trust is given upfront but needs to be verified. A team playing nicely and finishing the project reliably will most likely be hired again due to reputation, trust, and honesty.

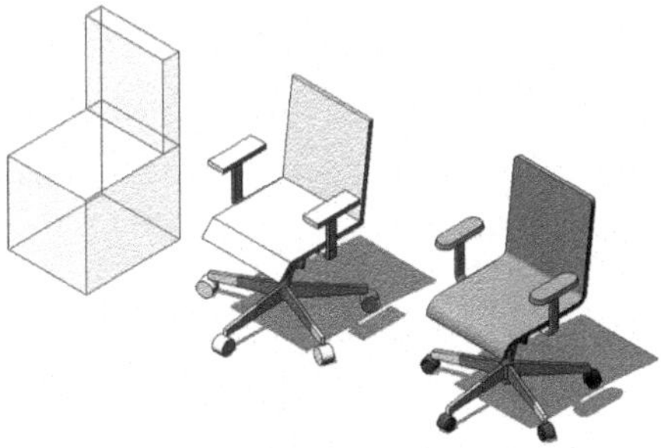

Fig. 6 Office chair example shown in different LOD

5 Leg 3: Technology

5.1 Level of Detail (LOD)

One of the most important success factors of integral approaches is the ability to adjust the level of detail (LOD) of the building elements during the different building phases. But adding too much data too early will slow down performance tremendously. It is crucial that all participants agree on the to-be-used LOD during each planning phase before the project starts (Fig. 6).

The '**C**'-Aspect:

Building professionals participating in planning and not being 100% sure how much detail is needed at the given moment tend to add the maximum available just to make sure that they do not forget anything. In early stages of design, when changes are still possible, it is mandatory that unnecessary data will not be added to the model as it would slow down the modeling process tremendously. The Set-in BIM manager or coordinator needs to watch over the LOD to make sure that too much additional information is taken out again. A BIM handbook, as mentioned earlier, is a very useful base to avoid endless discussions by defining this necessity for all participants before modeling starts.

5.2 Active, Passive & Exchange

Active data is a major advantage within the BIM planning phases. Building data can be set up as active parametric building elements that automatically adapt to changes in the design. The shown window example is one parametric building object placed into the BIM model with different parameter settings. So far, the level of parametric depth is not fully exchangeable between different BIM software packages.

Fig. 7 Same window element placed with different parameter settings

Exchanging active elements means to decide what version of the building element to use and loosing most of the active information. This step is permanent—the element becomes visible to everyone but is carrying just the information of the moment of export (Fig. 7).

The ‘one for all’, open source, free exchangeable BIM data exchange format is not yet in sight. The leading BIM software companies are currently still trying to establish their software as ‘the one and only’ on the market and are therefore not fully committed to develop a universal data exchange format. The ‘Industrial foundation class format’ (IFC), developed by the ‘building smart alliance’ is a good first step, but is still losing a great deal of active (parametric) model information during transitions.

The ‘**C**’-Aspect:

It is not uncommon to have someone just specialized in data exchange among the team who will keep continuous care of all ‘in and outgoing data’. Especially in larger projects in which planning partners are not using the same software, the needed data exchange can be so time consuming and therefore expensive that the whole project is unnecessarily slowed down. There are developments to force most of the project partners into the same software. Costs for licenses and software training can sometimes be lower on the long term than dealing with unsatisfying data import and export questions and the risk of losing important data. Changing software packages along the way is not recommended.

6 Leg 4: People

6.1 New Roles

Looking at BIM job descriptions worldwide the integral approach changed the demands for traditionally known roles in the building industry. The modeler, the coordinator and the manager emerged as new roles that include having a far broader knowledge and oversight regarding the ongoing overall process and the envisioned project goals (Fig. 8).

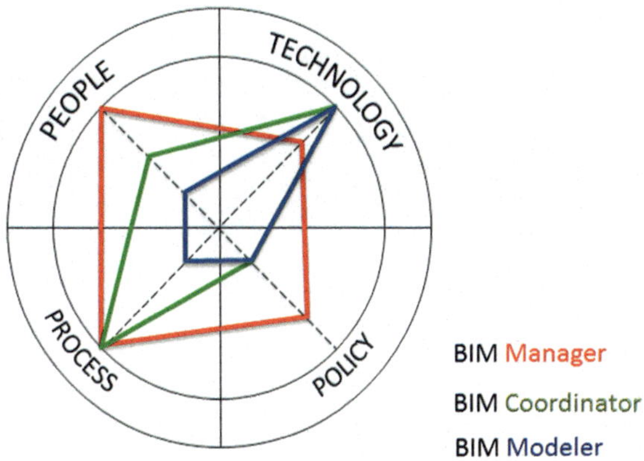

Fig. 8 New roles in BIM and their responsibilities

The '**C**'-Aspect:

To see your own role embedded in the whole building process is an important difference compared to the traditional building process where participants are able or even asked by their employer to hide behind their limited job responsibility and simply not intervening even if they see something that goes wrong.

As tasks within BIM are automatically connected to the development of building parts in time and process a broader understanding and education of all building professionals is required. The modeler especially needs to take and receive more responsibility and develop his or her process management skills to be able to play his or her part in BIM successfully. Schools, universities and companies are asked to adjust their education, curriculums, and training. Students today are still deeply connected to classical education and need to get their eyes opened to integral approaches.

6.2 *The Orchestra*

BIM requires far more complex ways of working together in teams than is common in traditional building. The traditionally known 'all overseeing Baumeister', still postulated by some, would have trouble keeping up with the complexity of needed decisions in a short period of time. New team working models, like being organized as an orchestra, seem to be a much more successful answer—as they make it possible to share the responsibilities with all participants involved (Fig. 9).

The '**C**'-Aspect:

A conductor does not know how to technically play each instrument but his or her role in organizing each participant right in time forms a masterpiece. To organize the process is the main part of the role of the BIM manager. He or she needs to know how to run the project in all its steps, yet the technical expertise is

Fig. 9 BIM manager tasks comparable to the ones of a conductor (*Modified on source* www.shutterstock.com, *#4408429*)

provided by others. The role of the manager can be taken on by the owner, the architect, the engineer or an external firm. As described earlier, it is important that the forming of the team is happening before the project start and needs to be written down in the BIM execution manual.

6.3 Error Culture

BIM thrives from making mistakes. This may sound odd, but making errors will usually make the project better. Mistakes in early phases of integral design are seen immediately by internal or external team members participating at the same moment. They can be easily corrected immediately and far ahead of the time that the concrete will flow (Fig. 10).

The '**C**'-Aspect:

A major social challenge will be to acquire a different mindset about accepting own faults as beneficial for the project. Building professionals who are traditionally educated and formed by existing building culture need to understand that when using BIM, faking and hiding will no longer be possible. The 24/7, equally shared visibility of all modeled building parts to everyone in the group means that there are always enough eyes watching and realizing what is going on in the model. Faking means that in all cases, someone will see it and point it out. Solving it later is possible but will mean more questions as the model is getting developed continuously and information will be added regularly.

What first may feel like being pushed or forced will soon be seen as a normal and quite effective way in solving challenges. Problems can be solved once they occur in the still digital model and not on the real site when changes and adjustments are costly and nerve wrecking. We all make mistakes and it is good to make them. Assuming that no one makes mistakes intentionally, they raise important

Fig. 10 Traditional error culture

questions which can be solved mostly without strong impact on the whole process in the moment that they occur.

7 Conclusions About the 'C' in BIM

In contrast to what glossy magazines and software developers often say—it is not the software that we have to buy and to learn to manage—it is the knowledge and the social skills we have to develop and the habits we have to change towards a better communication within the process of planning, building, and operations.

We have to accept that the demands of integral ways of working together are very different from what we did before on stacked paper or computers in 2D or 3D. We need to make the step from thinking in layers (imitation of stacked transparent paper) to object oriented building parts (and their continuous changing properties) over time within databases (Fig. 11).

Integrated approaches are possible only through working in teams where all team members meet at the same eye level. Everyone within the team is needed for BIM's success. BIM means to build virtually, but this virtuality has to be a digital twin of the building process and steps later on the construction site. To be able to feed the model, you need to know how to build in reality and to team up both internally and externally.

Internally, BIM requires the working together of young professionals who are skilled with computers, intermediate experienced professionals who are willing to organize, and experienced seniors who are motivated to provide answers to the most difficult questions.

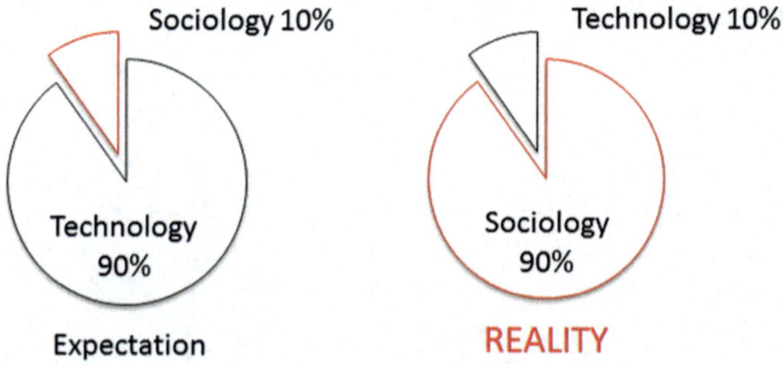

Fig. 11 The 90/10 rule. perception vs. reality (*Modified on Ref.* [4])

Externally, BIM requires the willingness of owners, authorities, planners, advisors and contractors to sit on one table and discuss in parallel processing what the projects needs in order to move towards a successful completion. The still very fast and active development of planning and building processes and technology will ask building professionals far more to engage in continuous learning and updating of skills and knowledge. Keeping up with BIM developments around the globe will be essential for building firms to survive and thrive in today's building markets.

References

1. Leslie, J.: The Trouble with Megaprojects, The New Yorker Online source: http://www.newyorker.com/news/news-desk/bertha-seattle-infrastructure-trouble-megaprojects (2015)
2. Rosenfield, K.: The Future of the Building Industry: BIM-BAM-BOOM! Online source: http://www.archdaily.com/262008/the-future-of-the-building-industry-bim-bam-boom/ (2012)
3. Egger, M., Hausknecht, K., Liebich, T., Przybylo, J.: BIM-Leitfaden für Deutschland. Forschungsinitiative ZukunftBAU, BMVBS (2013)
4. Deutsch, R.: BIM and integrated design. John Wiley & Sons Inc, New Jersey (2011)

Open Source Fabrication

Enrico Bassi

The potential of computational design and architecture seems to be limitless. It is a very common feeling in front of something so new and exciting that unfolds before us. The best part has probably yet to come, considering that it is still a pretty young discipline. I think that this is a tilting point in its evolution, mainly thanks to the parallel growing of new fabrication technologies. I will explain to you why I think they are so strongly interconnected.

Take a brilliant and ambitious project, designed by a genius before the development of a proper way to produce it, and you get the helicopter of Leonardo da Vinci. To actually get to an autonomous flying drone we had to wait for electronics, batteries, carbon fiber, etc.

Computational design needed a technology flexible enough to produce (almost) every possible solution out in an efficient and adaptable way. "Hand-made" as well as "industrially produced" can't be a suitable answer to such a dynamic way to generate projects.

What was missing was a cheap, reliable way to actually realize something directly out of a file; in two words, digital fabrication.

We can use this label to describe every way that produces something which involves a numerically controlled machine, in other words, something plugged to a computer that transforms the information, contained in a 3D or a 2D file, into instructions to produce an object or a part of it.

It doesn't mean it is always the best way to make something: if what I want is a simple geometry, that is easy to cast and which requires a huge amount of material, it is probably better to keep using traditional techniques.

On the other hand, some projects have complex geometry that does not come from a regular repetition of a standard module; with shapes almost impossible to describe completely with orthogonal views and too difficult to be made by hand.

This is the perfect matching point between computational design and digital fabrication (Fig. 1).

E. Bassi (✉)
Opendot FabLab, Milan, Italy
e-mail: enrico@enricobassi.com

M. Hemmerling and L. Cocchiarella (eds.), *Informed Architecture*,
DOI 10.1007/978-3-319-53135-9_6

Fig. 1 Opendot Fab Lab during Fab Academy, Milan

During the last six years, I worked as the coordinator of Fab Labs, fabrication laboratories which started at MIT as an experiment of professor Neil Gershenfeld. Quoting from Fab Foundation website.

> A Fab Lab is a technical prototyping platform for innovation and invention, providing stimulus for local entrepreneurship. A Fab Lab is also a platform for learning and innovation: a place to play, to create, to learn, to mentor, to invent. To be a Fab Lab means connecting to a global community of learners, educators, technologists, researchers, makers and innovators, a knowledge sharing network that spans more than 100 countries and 24 time zones. Because all Fab Labs share common tools and processes, the program is building a global network, a distributed laboratory for research and invention.
>
> A Fab Lab is comprised of off-the-shelf, industrial-grade fabrication and electronics tools, wrapped in open source software and programs written by researchers at MIT's Center for Bits & Atoms. Currently Fab Labs includes a laser cutter that makes 2D and 3D structures, 3D printers to make complex geometries, a sign cutter that plots in copper to make antennas and flex circuits, a high-resolution NC milling machine that makes circuit boards and precision parts, a large wood router for building furniture and housing, and a suite of electronic components and programming tools for low-cost, high-speed microcontrollers for on-site rapid circuit prototyping. Originally designed for communities as prototyping platforms for local entrepreneurship.
>
> Fab Labs are increasingly being adopted by schools as platforms for project-based, hands-on STEM education. Users learn by designing and creating objects of personal interest or import. Empowered by the experience of making something themselves, they both learn and mentor each other, gaining deep knowledge about the machines, the materials, the design process, and the engineering that goes into invention and innovation. In educational settings, rather than relying on a fixed curriculum, learning happens in an

> authentic, engaging, personal context, one in which students go through a cycle of imagination, design, prototyping, reflection, and iteration as they find solutions to challenges or bring their ideas to life.[1]

Because of its nature, a Fab Lab is the perfect place where computational design meets digital fabrication. The users of such a place are usually architects, designers, artists, as well as researchers, students, engineers, and developers. The mix of different skills allows the development of innovative projects, which are otherwise impossible.

On the other side, machines help to bring these ideas to life. There are a lot of interesting technologies, such as CNC milling and laser cutting, but probably, the most famous one is the 3D printing.

It is not necessarily the one with the bigger potential, and it is not always the cheaper choice (true only if you ignore the maximum dimension of the piece you can make). So why is it considered so revolutionary and disruptive? I'd like to think it is because so many projects are based on a worldwide community of makers and professionals sharing their knowledge for free. Today it is probably the larger existing community working on a single technology.

We will be back on this later, right now it's important to focus the attention on some basic features of 3D printing that matches perfectly with computational design.

In general, we could say that to 3D print something means to be able to solidify a precise amount of material in a specific point in space. There are of course a lot of ways to do so, but at the moment that's not so important. Just to bring two examples, let's talk about D-Shape and MX3D.

D-Shape[2] is a company founded by Enrico Dini, and works in aggregating together a powder to create very large stone-like pieces:

> The D-Shape printer enables full-size sandstone buildings to be made through our large-scale CNC machine.
>
> Existing materials such as reinforced concrete and masonry are expensive and inflexible for many complex geometries. D-Shape adds thin layers of binder to stone aggregates at predetermined slices, this gradual build-up of layers allows complex and internal geometries to be fabricated easily. The design process remains similar to existing 3D printing technologies, now possible on a larger construction scale.
>
> D-Shape has been designed to compete with cement, reinforced concrete, bricks and stone, now focusing on digital design and natural ecological materials.
>
> The machine consists of a rigid 4 × 4 m frame, a large flat bed and a custom print head holding up to 300 nozzles. CAD-CAM software drives the machinery during the building process, scanning across the surface and depositing droplets (voxels) of binder.

A completely different approach is used by MX3D[3] (Fig. 2).

---

[1]http://www.fabfoundation.org/fab-labs/what-is-a-fab-lab/.

[2]http://www.d-shape.com.

[3]http://www.mx3d.com/.

Fig. 2 A large format D-Shape machine, photo by Enrico Dini

> We believe that we are at the beginning of a revolution: digital design and fabrication are changing rapidly. The possibilities are endless. MX3D is a highly innovative company that developed a groundbreaking additive manufacturing method. In 2014 we invented an affordable multiple axis 3D printing tool. We equipped an industrial robot with an advanced welding machine and developed software to control it. We can 3D print metals and also resin in mid-air, without the need for support structures. With our MX3D Bridge project we will showcase how digital fabrication is finally entering the world of large scale, functional objects made of durable materials.
>
> In order to innovate, we research and develop our technique continuously. For this we found the perfect R&D partners and sponsors. Together we are ready to 3D print the first fully functional steel bridge in the world to put our revolutionary technology to the test. We offer creative robotic manufacturing solutions that we believe will transform Arts and Industry. Digital manufacturing like this is still unexplored territory and leads to a new form language, an aesthetic freedom. And it might transform processes at construction sites as they are now.

Both projects base their building process on 3D printing, but the solution they adopt is unique and created just for this scope.

Of course this opens opportunities otherwise almost impossible if limited to existing, out of the box, commercial machines. Even if the solution you need to use is unique and must be developed "ad hoc", having access to a wide range of open source projects speeds up the development of innovative offspring a lot. We don't need to start from scratch every time, we don't need to reinvent the wheel for every travel. "Open source is all about don't have to solve the same problem twice" says Alastair Parvin, one of the Wiki House project's founder.

To develop something innovative, that you can actually make, you don't necessarily have to build your own tools. A lot of projects were made using existing, affordable tools. One of the first examples, at least in Europe, is the "Fablab House" (Fig. 3).[4]

> The Fab Lab House is a self-sufficient dwelling produced to take part in the Solar Decathlon Europe 2010 competition by a consortium of organizations and companies led by the Institute for Advanced Architecture of Catalonia, The Center for Bits and Atoms at MIT and the Global Fab Lab Network. The objective is to design an integral solar house with the technologies of our time, which will generate maximum resources with minimum investment. A house built for people, committed to creating the city, and connected with the whole world. The Fab Lab House is not simply a box with solar panels on its roof; its physical structure is integrated with its energy production and management of the information it generates. The Fab Lab House, developed with an open design, can be fabricated with local materials anywhere in the world. Its organic form, which responds directly to its environment, can be produced using advanced industrial systems which allow each dwelling to be made to measure for its users. If the twentieth century championed the premise that 'form follows function', in the 21st century 'form follows energy.' The house is no longer a machine but an organism for living in. The Fab Lab House uses the resources of its environment—sun, water and wind—to create a microclimate that passively optimizes the basic conditions of habitability.

The capability to design something that is a direct consequence of an algorithm finally allows us to break down the complexity of the projects into its elements, into blocks or functions that compute each single aspect, starting from an input and influencing the final project as an output: besides the shapes, you could calculate solutions based on latitude, average temperature during the year, landscape, apparent path of the sun, and so on.

If each block would be written properly, some pieces of the project could be reused, integrated, modified, and shared with others.

It's not necessary anymore to be able to develop everything by yourself; the cooperation between creative people around the world finally has the common language it needed: computational design. Just to bring you an example, the network of Fab Labs counts over 1000 labs (www.fablabs.io/labs). Even if the labs have some differences, it's safe to assume that in each one of these labs, you have access to a set of similar technologies, and as a consequence, capability to produce the same piece (or even better, a modified version) of the original one.

To fully get access to the potential of a network, of course, we have to let the knowledge spread freely. Few years ago, it seemed that open source software was just a niche for nerdy users and stubborn independents. Now it's not like that anymore, we commonly use pieces of code or open source software to actually work, study or bring on our ideas.

The same trend seems to apply to design and architecture. Of course that make sense when the projects can be easily manipulated, modified and adapted, or if different people can develop and share their work in an effective way, for instance,

[4]http://www.fablabhouse.com.

Fig. 3 Fablab House, picture from Adrià Goula

think about Grasshopper,[5] the most famous plug-in for Rhino to do parametric, generative and computational design.

Every time someone develops and releases a new function, everybody in the community benefits from his work including the developer himself who will probably receive support for his job from others.

A code is not a box of cookies: sharing it with others is not limiting the access you have for yourself.

Actually it is a great way to build reputation, start collaborations and to prove what you are capable of. This does not apply only to lines of programming languages, we can share much more than that.

Wikihouse, for instance, relays on a large community of people working around the same idea, hoping to develop a sort of standard that everyone can use as a starting point, both to build a house out of the box or to develop new solutions.

It is interesting to see what are their "Design Principles".[6]

- ***Share global, manufacture local***
  "It is easier to ship recipes than cakes and biscuits" John Maynard Keynes.
- ***'Be lazy like a fox'***
  Don't keep reinventing the wheel. Take something that works, copy, adapt, give credit and share. (Thanks Linus Torvalds & Eric S Raymond)

[5]http://www.grasshopper3d.com/.

[6]http://www.wikihouse.cc/about/.

- ***Open materials***
  Design for cheap, abundant, standardized, sustainable and, if possible, 'circular' materials.
- ***Start somewhere***
  You can't solve everyone's problems in one go. Design something useful for a context you know and understand, then share so others can adapt to their economy, climate and context. Release small, iterate and 'fork'.
- ***Higher performance, lower thresholds***
  Design to lower thresholds of time, cost, skill, energy & resources in manufacture, assembly and use.
- ***Open standards***
  Share and make shareable. Where possible, work to existing open standards.
- ***Safe***
  Maximise the safety, security, health & wellbeing (physical & mental) of users at all stages of a product's life.
- ***Inclusive***
  Look for ways in which age, gender or disability might be barriers, and try to design them out. Design products, processes and documents that are accessible and intuitive.
- ***Modular***
  Design hardware and software that is interoperable, product-agnostic and flexible, so elements can be independently altered, substituted, mended or improved.
- ***Design for mistakes***
  Make it hard to get wrong, or not matter if you do. The Japanese even have a term (Ed. It is "pokayoke") for this.
- ***Design for the next Normal***
  Design beautiful, high-quality products that lower cultural barriers and make radically sustainable, sociable design 'normal', rather than 'alternative' or 'fashionable'.
- ***Knowledge should always be free***
  But professionals' time should be paid for.
- ***Superpower citizens***
  Afford as much capability and choice to citizens as practically possible. Democracy is a good design principle.
- ***Neutrality***
  All companies can participate in the WikiHouse commons, but no one company ever gets a monopoly or lock-in.

The suggestions given are not just good design principles; these are the borderline and the manifesto of the project. If the objective is clear to everybody, it is easier to decide to join in, participate and develop something useful to the entire network. In some ways, it happened even before the use of computational design but it is easier to fork, edit and manipulate data and code than lines and shapes and pieces of paper.

The key point when we talk about community working together is to allow everyone to stay focused on what they are good at. Let's keep in mind that "*Everyone is a genius. But if you judge a fish by its ability to climb a tree, it will live its whole life believing that it is stupid.*" The objective here is to allow everyone to express his/her full potential, and enjoy what he/she is doing, being a part of something bigger. By nature, designers and architects are usually very good at thinking about the whole picture, the big vision that keeps all the pieces together. To do so, you can't go through all the small technical details needed. A similar thing happens in a factory or any other complex system: everyone has a specific role that plays a part in the entire process. Overviewing the entire process is a specific role as well; in the case you can't go into particular processes, the risk is to get stuck there and loose the capability to stay focused on the result.

In a shared project, everybody naturally does the part of the process they like the most: it could be writing a tool (like Grasshopper), developing a plug-in (like Millipede), figuring out an interesting way to use it [for instance how to optimize the roof structure with millipede or to do a specific project (the actual design of a specific building)]. Computational design improved this process a lot, helping to share both ideas and tools to develop it without creating "clones" of an original work.

The scalability of this approach is another interesting point to talk about: the main obstacle to solve is often the absence of a proper tool, a block to compute a solution that works. A single wrench could fix a thousand cars, in the same way, a piece of code could be used in thousands of projects, without being repetitive. Adjusting a pattern to the curvature of the surface could be a part of a tessellation, as well as a mechanical optimization. The "core" is the same, but the final result is completely different.

Until a few years ago, an architect was often following almost the entire development, in particular, for small projects. If you think about it, it is a pretty big limitation that could strongly reduce how innovative or experimental the project could be. It is like a do-it-yourself process; all the steps from the seed to the flour, just to bake bread. If a set of solutions, which is like a sort of toolbox, is available for everybody, it lowers the barriers to access a specific process, and that frees a lot of energy to invest in looking for solutions. By nature, creative minds that are sensitive to problem solving, are objective driven. They are not usually interested in hardcore coding, hence, modern computational tools offer a way to free them from struggling with technical issues, to keep their mind on searching for new things, analyzing the changes, and looking for a glimpse of future.

So what I'm saying is that we could think about the community, that contribute to making computational design interesting and accessible to everyone, as a global group that works both together (on the development and diffusion of the tools) and singularly (on the personal, creative project). It is a more complex contest, compared to what we are used to, but it is worth the effort to face the change.

Now, computational design is not all about shapes and technologies to produce them, it is also a deep change in our mindset and it is a new way to face problems and their solutions. In some ways, it is to design like nature does: creating a code, that generates a product, influenced by the environment and adjusted to that.

There are two levels of adaptability: one is not implying a change in the code while the other does. If you take two seeds from the same tree and you plant them in different places, what you will get are two pretty different results. The depth of the ground, the amount of rain, the direction of the wind, the intensity of the light and so on, will shape them in different final results even if the "instructions", the DNA, would have been the same. It means we could use the same design and let the environment shape it in all the possible inclinations: just input different data each time to get countless alternatives. Even when repeating the same project and the same way to produce it, we won't repeat the same product.

Now, a plant that adjusts physically to the environment does not involve a mutation of "the code", it is just a reaction to the external conditions. On the other hand, evolution does imply that mutation, and following this path, is the creation of something new.

It seems that camels are actually arctic creatures: the large paws are not "designed" for the desert, but for the snow, the hump is a tank of fat, to survive during the long and cold winters. Migrations forced these features to fit to a new environment, and now, to almost everyone, a camel is a desert creature. If we go on with this analogy, we could think about it as a collection of pieces of code, a program written with blocks that describes all the characteristics of the animal. What if we could take away the "store fat" piece to increase the speed or combine all these blocks together to create animals that does not exist, yet, but perfectly responding to some specific function or territory? It would be a sort of "smart evolution", but it will be much simpler to do than having to write the entire DNA from scratch. Nobody wants to develop GMOs of course, but I am sure that a lot of people would enjoy playing with a simulator based on "animal function blocks" to see the result, while the idea of writing a very long sequence of "A" "T" "C" "G" would be less attractive.

Getting back to the code, each "block" we were talking about, is a way to define a function. Once someone has written a piece of code that will allow to do something based on a parameter (for instance the dimension\pattern of the brise soleil based on the apparent path of the sun), if it is shared in open source, it could be used in different projects, freeing the architect from a technical issue; less hassle, more solutions.

Think about how many repetitive, complex, delicate, sensitive optimization or calculations you have to apply to your project. All these things are parts you could legitimately automate using code. If it is true, as it is, that every building should be integrated and influenced by the surrounding environment, computation is the ultimate tool to do so.

Other important areas developed in Fab Labs are electronic and embedded programming. Just to report the simplest example of all: writing a simple piece of code, that when uploaded in an Arduino, will make an object interactive.

Arduino is probably the most famous open source, easy to use microcontroller (arduino.cc):

> Arduino is an open-source prototyping platform based on easy-to-use hardware and software. Arduino boards are able to read inputs—light on a sensor, a finger on a button, or a Twitter message—and turn it into an output—activating a motor, turning on an LED, publishing something online. You can tell your board what to do by sending a set of instructions to the microcontroller on the board. To do so you use the Arduino programming language (based on Wiring), and the Arduino Software (IDE), based on Processing.
>
> Over the years Arduino has been the brain of thousands of projects, from everyday objects to complex scientific instruments. A worldwide community of makers—students, hobbyists, artists, programmers, and professionals—has gathered around this open-source platform, their contributions have added up to an incredible amount of accessible knowledge that can be of great help to novices and experts alike.

When you think about it, computational design allows you to create incredible three-dimensional shapes, based on the environment, algorithms, data, and patterns. So, what if you could let the final result interact with the external input and react to it? What if you compute the behavior of a dynamic object, not just a static one? Soon we will probably have to introduce the concept of a 4D file: 3D data, plus the code that animate its behavior.

This "forth dimension", the electronic + the embedded code, seems to be put to use only after the object is already built. It is also possible to flip the order of the steps and use sensors to actually collect the info that will be the starting point of the project, where, instead of using them to animate the final result, its function is to collect data and process them. For instance, with Firefly (www.fireflyexperiments.com) it is possible to communicate info between Grasshopper and Arduino to use, as input, online data, environmental parameters, or sensors feedback.

These sensors could read the stresses and forces applied to the object, adjust the 3D file in a dynamic way, optimizing the distribution of material, etc. Even if in this case it is more complex, it is basically what you do with a topological optimization. Every architect and engineer could easily calculate the dimension of a pillar, if the weight it must resist to is known; it is actually very simple math. When the shapes are more articulated, not perpendicular, nor repetitive, or stressed by different forces (beside the weight of the structure itself), it gets less trivial, sometimes almost impossible to calculate.

It is not just the structural integrity that could be complex to calculate: sometime the resistance of a frame, the insulation of a specific pattern, the feasibility and the time needed to make a particular object in 3D printing, is simply something easier to do and test rather than to simulate or calculate; digital fabrication, again, is the best way to do so. Generating variants, testing them and keeping the most effective is basically what evolution does: it is not by chance that this particular way to do computational design is called "genetic design". The process is simple: starting from a solution that seems reasonable, a certain number of variations are created. Using a digital technology (i.e. 3D printing) they are quickly and precisely created and each piece is tested to check which is the best one. The "survivor" is the starting point of the next iteration. In theory, the process could be repeated a countless number of times, but as soon as the variations are not improving significantly compared to the performances of the previous ones, it makes no sense to go on.

As we have said, the possibilities of computational design are far beyond the capability to make "weird" shapes. We need to use it to evolve the way we design, to go over the standardized set of materials, tool, processes, way to test and certify, and so on. Because of this, computational design could be the only available solution to get your goal.

It is the case of WASP Project (www.wasproject.it/w/en/maker-economy/), an Italian company that produce and sell very well made 3D printers. Their vision is to develop a "maker economy":

> Maker economy is a new model where everything can be produced by yourself, where there is the chance of not depending on some insurmountable entity that holds the productive monopoly. WASP is the acronym of World's Advanced Saving Project and our research is focused on the common welfare and shared knowledge. It is not necessary to be great to have great topics, our big goals are the sense of our work. House, work and food, health, are what a man needs to live. [...]
>
> 3D printing allows to find some new shared solutions for each one of these topics. The Maker Economy model can be exported everywhere, but, more than in our society, it could work where it doesn't exist a productive fabric and a net of working infrastructures. It is a sustainable development that feeds itself and doesn't cause logistic problems. [...]
>
> We need everyone's contribution: architects, structural engineers, green building and renewable energy experts. Together we are going to project not only a house but a real system where the self-production will become base of a new Economy. A self-sufficient city where sharing knowledge and to give equal opportunities are the common denominator.

Fig. 4 The 12 meter high Big Delta form Wasp, designed to build houses. picture from Wasp Team

Their main goal is to evolve the 3D print technology up to a level where it could make a house. To actually do that, using local materials (the process is inspired by the mud architectures), it is fundamental to adjust the structure to the material, to figure out a way to guarantee the safety, to generate the instructions for the 3D printer itself (Fig. 4).

I think it is not a coincidence that in the description of their work, they underline, so strongly, the importance of the collaboration and the community.

Hence, computational design has a brilliant future in front of it, full of unfolded opportunities, and it could change a lot of aspects of our way to approach problems and to look for solutions. I am positive that it will happen only if it will evolve, together with the technologies, our way to think as well as if we will learn how to cooperate.

Robots in Architecture, Research and Development

Pierpaolo Ruttico

Thanks to the growing awareness of the integration of design, material, and production process, and thanks to increasingly widespread digital technologies, new perspectives for the construction industry and new opportunities for designers and system integrators are emerging. The key players of this scenario are industrial robots, manipulators that move along programmed paths and process any material with precision and speed. After ten years of research into robotic fabrication and automation in the field of architecture, the aesthetic and functional potential generated is transforming architectural design and construction culture on a grand scale. This potential emerges clearly when the elements that make up architecture generate patterns where variation plays an important role in defining gradients of transformation, whether it be of shape, structure, or performance in the energy field (Figs. 1 and 2).

Thanks to the use of mathematical models to manage variations, it is possible to generate structures in which the whole is more than the sum of the individual parts. This leads to the design of buildings and objects that vaunt new expressive connotations, so the time has now come to design competitive production systems for delivering consistently differentiated elements. The cutting-edge aspect of robotics applicable to architecture is the serial production of nonstandard, bespoke elements —known as mass customization—using automated processing and assembly techniques. The industrial robot is flexible and adaptable by nature and can perform repetitive operations using different tools (called end-effectors), managing tasks with extreme precision and speed; a range of action of at least six degrees means it can perform complex, manifold manipulations and manufacturing processes.

P. Ruttico (✉)
IndexLab, Politecnico Di Milano, Lecco, Italy
e-mail: pierpaolo.ruttico@polimi.it

M. Hemmerling and L. Cocchiarella (eds.), *Informed Architecture*,
DOI 10.1007/978-3-319-53135-9_7

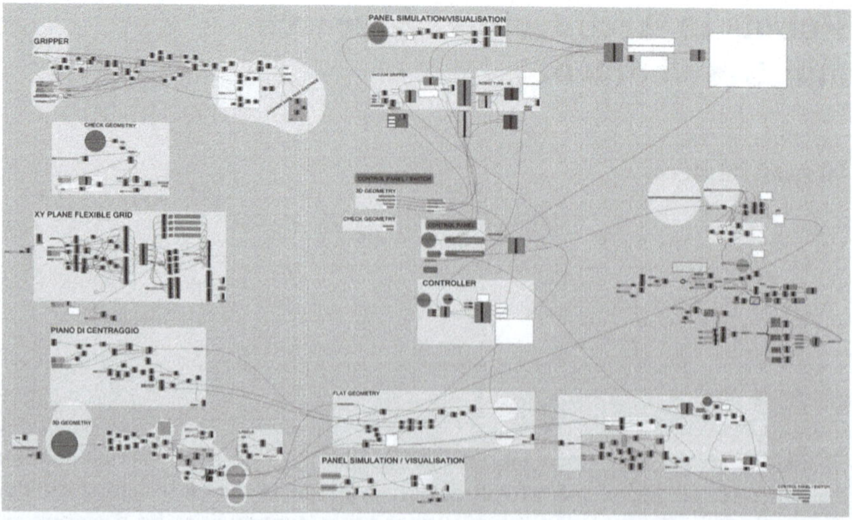

Fig. 1 Algorithmic design platform canvas—Grasshopper for Rhino. Indexlab 2010, Politecnico di Milano

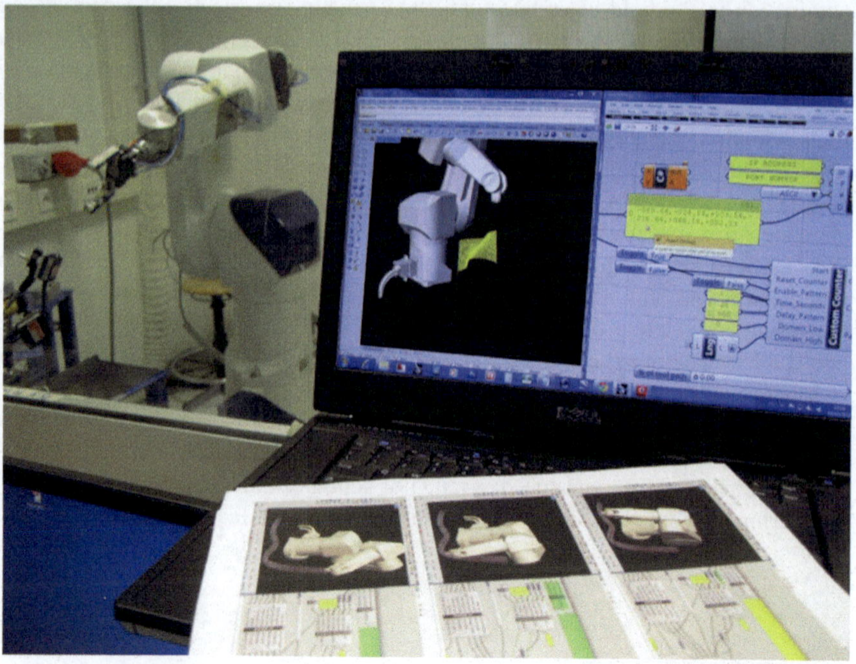

Fig. 2 Custom-made script to control a six axis anthropomorphic robot. Indexlab 2011, Politecnico di Milano

The experiments conducted make the best use of robot performance while investigating the characteristics of different materials, evolving a new way of conceiving the processes and opening up new design methods, where form is not arbitrary, but linked to the production method. The focus is not so much on designing forms as a priority as compared to defining rules that use algorithms to govern construction processes operated by robots. The fundamental contribution in this area, deriving from university research over the past decade, was the invention of adaptive and flexible systems that facilitate the design and the manufacturing process of components used in building construction, combining serial production with personalization. The goal is to combine form and performance in different components using the intrinsic properties and mutual efficiency of different materials: an engineering process that uses computational design and heralds a sustainable built environment, with a greater degree of functionality and integration between the products and the environment. These new design and production techniques can expedite the creative process and produce architecture that is totally coherent for concept and construction.

In the scientific community, Gramazio & Kohler is a name synonymous with robotics applied to architecture. At ETH Zurich, Fabio Gramazio and Matthias Kohler founded the first digital fabrication laboratory in 2005. Their research has influenced dozens of universities and hundreds of researchers and in addition, they have pursued several experiments, from oriented repetitive module assembly following predefined patterns to the use of fluid materials.

An important step towards the development of robotic fabrication research was made by Sigrid Brell-Cokcan and Johannes Braumann. In 2010 they founded what was initially a TU Wien spinoff: The International Association for Robots in Architecture. The goal was to build a network of research laboratories and to spread the culture of robotic fabrication. In 2011, they released KUKA|prc, a plug-in for Grasshopper, which enabled robot control directly within CAD for the first time.

In France, Philippe Morel was one of the first theorists and innovators to have sensed the potential of robots and his EZCT architecture firm, co-founded with Felix Agid and Jelle Feringa, was a pioneer in applying robots to research. In 2010, it was precisely at EZCT that Thibault Schwartz began to write HAL control software. The next year, the software was launched and contributed to the growing use of robots by students, architects and designers. During the same period, Jelle Feringa founded a laboratory in the Netherlands at Rotterdam's RDM campus as part of a Ph.D. project at the TU Delft—Hyperbody lab. Its contributions were significant: in 2012 he produced an application with diamond wire for cutting concrete and marble, and within the following year, this led to important industrial developments. His research team is currently engaged in trials of robotic three-dimensional printing of concrete. Another important reference in the scientific community is Achim Menges, whose successful experience at the London

Architectural Association was followed, in 2008, by the foundation of the Stuttgart Institute of Computational Design (ICD). Here, each year since 2010, his research team has created a research pavilion using state-of-the-art design and process techniques using different materials (from wood to composite materials with carbon fiber), exploring their characteristics and potential. In Italy, the research proposed and developed by Indexlab at Milan Polytechnic, with support of the departments of mechanical engineering and of architecture, built environment and construction engineering, together with the contribution of private companies, is part of this scientific-operational community. Indexlab presented original contributions starting from "Advances in Architectural Geometry 2012" (Bitmap-driven parametric wall for robotic fabrication, Fig. 3). Thanks to the interest and support of partner companies, the research at Indexlab has grown fast, pioneering flexible and automated production processes which are more competitive than current—state-of-the-art—manufacturing methods. Robotic assembly for free-form structures and claddings using robotic placement of differentiated flat panels with adaptive attachment is just one example (Figs. 4 and 5). Equally significant is the system that generates new applications using robotic reconfigurable mold production processes along with thermoforming processes (Figs. 6, 7, 8 and 9).

In the United States, testing began and developed in the same way as in Europe, with constant contact and swapping of information with European researchers. Wes McGee and Brandon Clifford initiated the robotic fabrication experiments at Taubman college and hosted Rob/Arch 2014 (conference on Robotics in Architecture) at the University of Michigan. The 2016 edition of Rob/Arch moved to Australia and teamed up with prominent Australian Universities to co-chair international workshops: RMIT, Monash University, Bond University, UNSW, and UTS. Beyond Rob/Arch, there were, and there still are, various opportunities for exchange and mutual growth through international conferences and workshops on digital and robotic fabrication, such as Smartgeometry, Advances in Architectural Geometry, Fabricate, and ACADIA.

Academic research coupled with new industrial technologies in this area are spawning new companies, where design and fabrication occur simultaneously. There are several emerging start-ups committed to innovating production methods for serial repetition. In Switzerland, for example, ROB Technologies deals with engineering of non-standard applications for the assembly of brick blocks, wood, and ceramic elements. In Denmark, ODICO Formwork Robotics applies robotic cutting to molds for production of individual concrete elements (Jelle Feringa is now Chief Technology Officer within Odico). In France, XTreeE (ex-EZCT team and ex-students from Philippe Morel and Thibault Schwartz) is the first company to commercialize robotic large-scale 3D printing technology for the architectural design, engineering and construction sector. These are communities committed to seeking new applications, new interfaces, and management software (Figs. 10, 11 and 12).

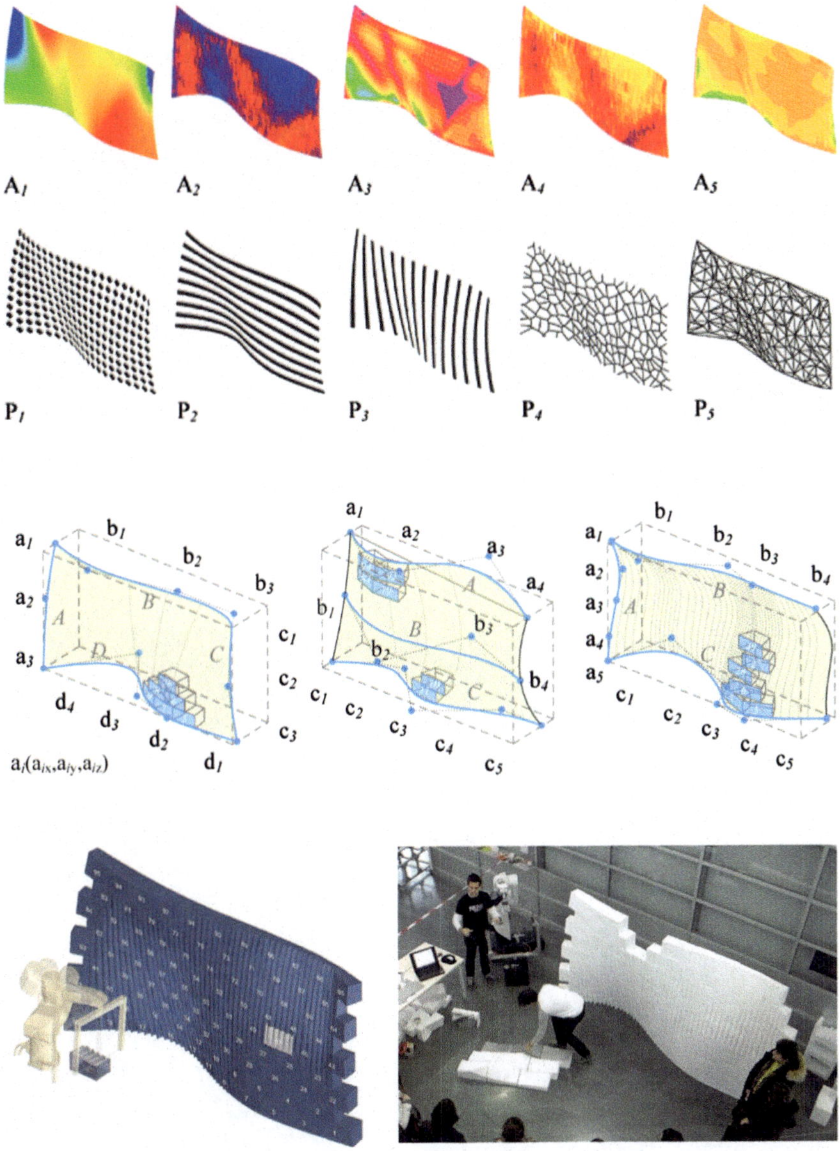

Fig. 3 Bitmap-driven parametric wall for robotic fabrication. Indexlab 2012, Politecnico di Milano

Fig. 4 Circle packing. Robotic pick and place and adaptive welding. Indexlab 2013, Politecnico di Milano, Nieder

Fig. 5 Circle packing façade. Indexlab 2013, Politecnico di Milano, Nieder

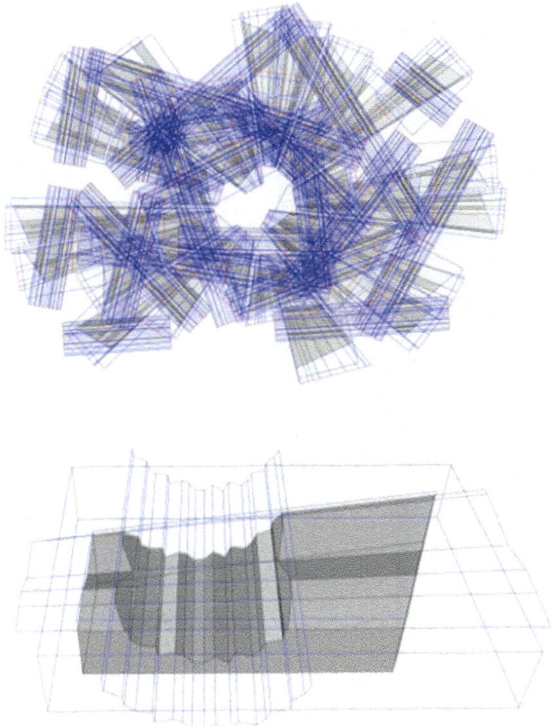

Fig. 6 Design of rulings for robotic hot-wire cutting. Indexlab 2014, Politecnico di Milano

Fig. 7 Robotic hot-wire cutting and thermoforming. Indexlab 2014, Politecnico di Milano

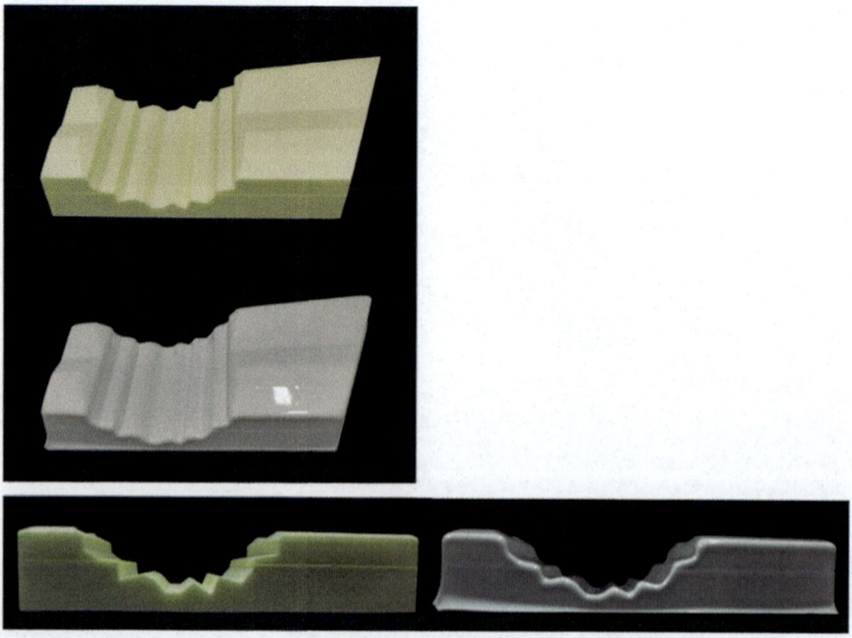

Fig. 8 Photos of a foam mold (*left*), thermoformed with a PVC sheet (*right*). Indexlab 2014, Politecnico di Milano

Inventing new production approaches and delving into corresponding design approaches allows architects, designer and engineers to raise awareness of “doing it the digital way”.

The use of robotic fabrication in architecture and construction has grown rapidly over the last decade, and continues to accelerate as the potential for innovation and

Fig. 9 Panel results over different geometries. Indexlab 2014, Politecnico di Milano

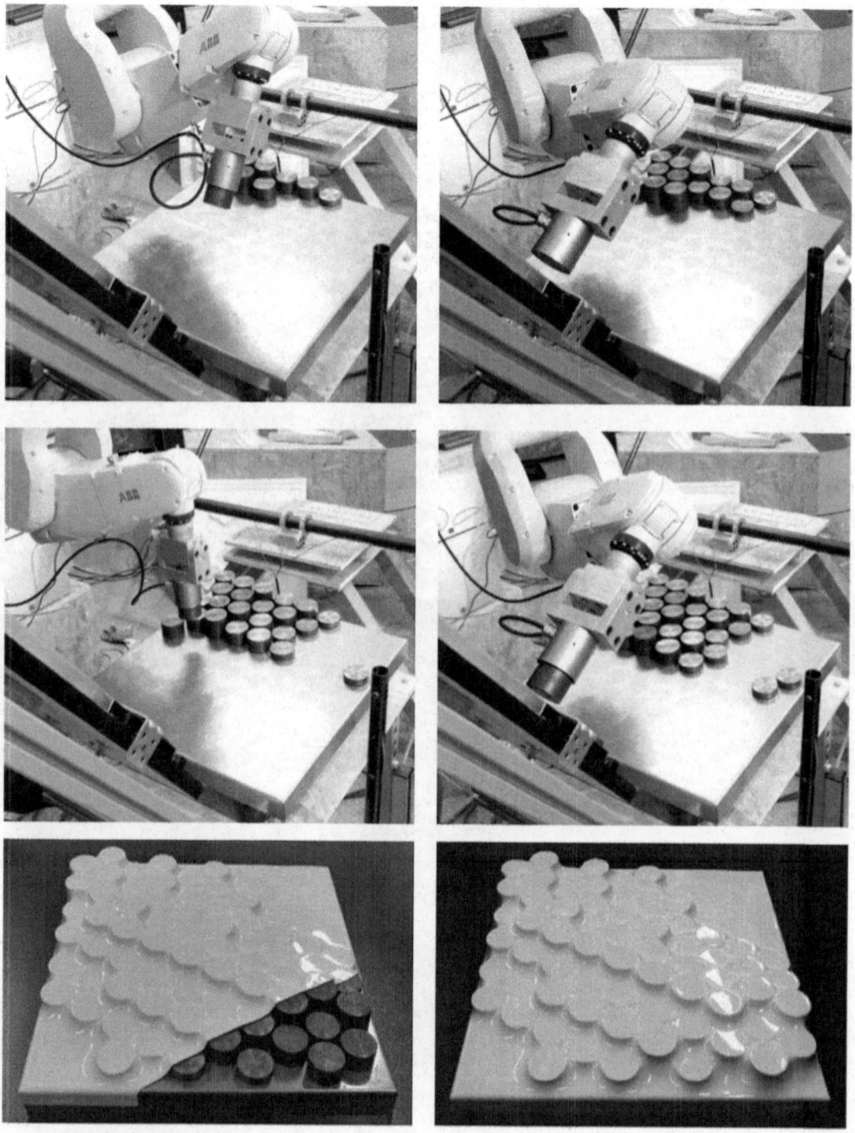

Fig. 10 Robotic pick-and-place of reconfigurable mold elements: images of the process. Indexlab 2015, Politecnico di Milano

Fig. 11 Setup for concrete casting over the generated molds (*left*); Final composition (*right*). Indexlab 2015, Politecnico di Milano

Fig. 12 Delaunay façade. Robotic adaptive assembly. Indexlab 2017 (Pierpaolo Ruttico)—Mecc (Francesco Braghin), Politecnico di Milano

creativity using robots is harnessed by the creative industries. Beyond setting protocols and automation processes for fabrication, we are expanding the possibilities of human-machine interactions for collaboration in industry. This has the real potential to lift economic growth for the building construction industries.

Mass Customization in the Era of Industry 4.0: Towards Immaterial Building Technology

Ingrid Paoletti

1 Introduction

Digitalization of manufacturing in the era of Industry 4.0 is slowly but strongly challenging also the construction sector. The possibility to introduce Mass Customization of products is possible thanks to two very important factors: *computational design process* and new *advanced manufacturing machines*.

These two advancements can influence the construction industry, towards 4.0, pushing to deliver not only customized products for special projects but also to increase the quality of building construction in general with techniques that can be designed ad hoc and personalized in material and technology with a cost near to standard production (i.e. Mass Customization).

The amount of information that can be incorporated in a building system can foster a sort of Immaterial Building Technology, where performances are embedded in the early design thanks to computational techniques, manufacturing opportunities and innovative workflows.

2 Digitalization of Manufacturing in the Era of Industry 4.0

Globally there is a high awareness that the society stands on the brink of a new industrial revolution, driven by technological breakthroughs such as advanced computing, big data analytics, robotics and manufacturing.

I. Paoletti (✉)
ABC Deptartment, Politecnico di Milano, Milan, Italy
e-mail: Ingrid.paoletti@polimi.it

M. Hemmerling and L. Cocchiarella (eds.), *Informed Architecture*,
DOI 10.1007/978-3-319-53135-9_8

These technologies open new horizons for the industry to become more efficient, to improve processes, and to develop innovative products and services. They also help the industry to respond to the customers demand for personalised products & services, safety and comfort as well as improved energy and resource efficiency.

Many researchers have already named it Industry 4.0, tackling it as the revolution that refers to the possibility to manufacture with a very high quantity of data.

Schaub defines Industry 4.0 as the era of Cyber Physical Systems, defined as an advanced tool creating a virtual copy of the physical world that allows to anticipate decision. Moreover, human interaction becomes a key point of this new model of Industry, enabling the value chain to increase with machines that can communicate with each other and also to become easier and more safely with humans.

European industry is strong in sectors such as automotive electronics, security and energy markets, manufacturing, robotics, telecom equipment, business software, laser and sensor technologies but sometimes seem to forget how these applications can be fostered to construction industry, which in reality has a lot of responsibility in society development.

The main reason is that even if Europe also hosts world-class research and technology institutes, many parts of the economy have been quick to take up digital technologies and processes-like high-tech sectors-, while construction face strong competition from other parts of the world and, unfortunately, many traditional industries are small and medium enterprises (SMEs).

SMEs have a higher difficulty to lag behind due to the high cost of access to high levels of information, research and development, cutting edge technologies.

It is true however that a strong wave of experimentation is starting from universities and research centres fostering the innovation that the sector does not seem to get, yet, from an industrial point of view (Fig. 1).

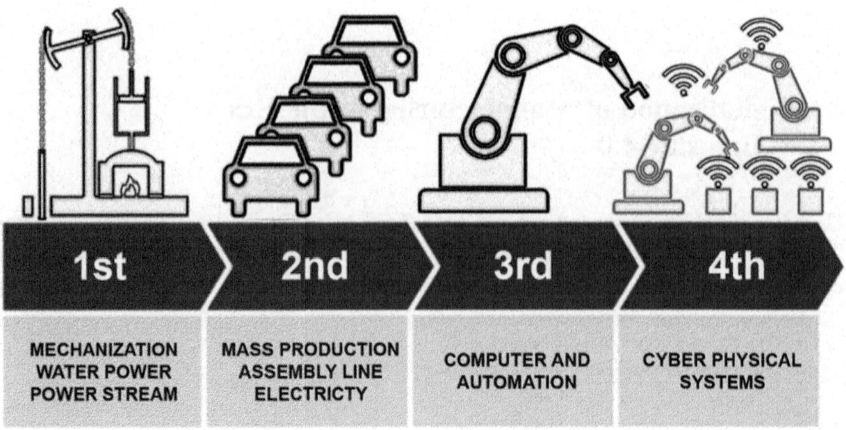

Fig. 1 Industry 4.0—evolution schemes from first to forth industrial revolution (Modified on Christoph Roser at AllAboutLean.com)

3 Mass Customization for Construction Industry: Computational Design and Advanced Manufacturing Techniques

Mass Customization is a term used from nineties by B. Joseph Pine which has, for a long time, referred more to a business model than to a real advancement in the process.

Mass Customization refers to the possibility to derive new elements from existing products that can be personalized thanks to innovative tools, new machines, and advanced processes, while avoiding the high cost of personalization.

Manufacturing techniques and innovative production methods in construction seem often quite resilient to change. This is due to traditional construction methods and consolidated process of production where innovation is often a very slow process, driven by economic reasons more than by the need of effective new products or systems.

Today advanced construction processes are driven by novel design procedures, among them computational design, early stage engineering, optimization of materials and systems which enable new ways of manufacturing [1].

This means that a new way of conceiving mass customization can be determined by instruments and tools enhanced by workflow information, keeping it more near to knowledge than only in lean production, strategies whereby a production strategy is focused on the broad provision of products and services [3].

Providing individually customized products by using flexible computer-aid process and organization structure with reasonable low costs and lead-time will also help the integration of personalized products into a traditional process and their physical assembly.

Compared with mass production, a wide range of combinations of product features may result in innumerable variants for a single product, which makes the number of product variants (i.e., product variety) increase drastically. Product family design, recognized as an effective means to support product variety with minimal data redundancy, has become one of the prevailing approaches in implementing Building Technology [6].

Mass customization puts at the centre the idea that customizing industrial production is gaining a higher relevance either in the design concept or in the production and construction phase.

Two important technologies are allowing this innovative Mass Customization: computational design and advanced manufacturing techniques.

3.1 Computational Design

Computational design refers to a procedure which interprets mathematical data with the idea that a process can be informed in order to give directional results.

The computational method can act on at least three sides. One is the design process, where it is possible to engineer a specific form with performances while drawing it from the early concept. The second one is the possibility to customize machines to fabricate specific machinery for the materialization of a design.

The third one is, through the two described before, to engineer specific custom-made systems with specific embedded properties for production.

Depending on the specific use of the building system, each designer can decide to activate certain properties more than others and the embedded performances of the material become active drivers of the design [2].

The advancement in this possibility to 'compute' and therefore to add information while developing a specific product, widens the opportunities to create new object that can be optimized in shape, material and production methods.

In Architecture and Construction, this is, really, an enhancement of the possibility to conceive and create innovative ways of building a system or a component. Its concern lies with the process of dealing with information, thus increasing the level of accuracy of the project (Fig. 2).

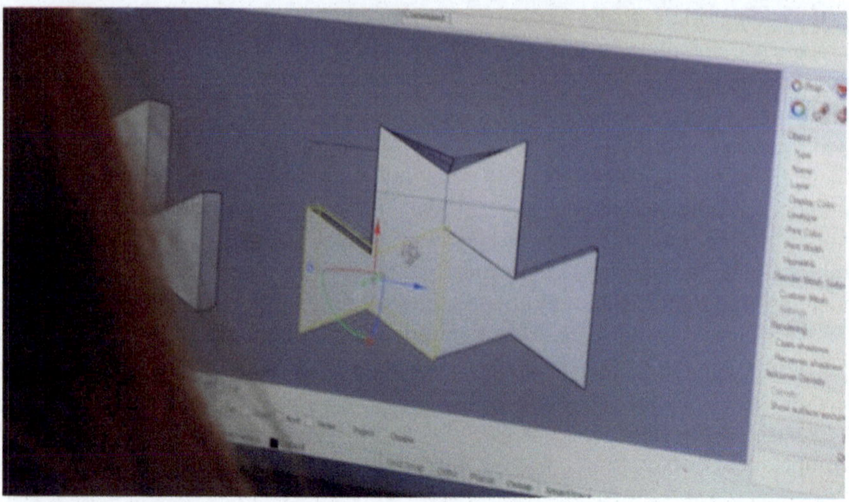

Fig. 2 Interlocking system of polymeric components designed with specify rules for assembly and production (ACTLAB, ABC dept., Politecnico di Milano)

3.2 Advanced Manufacturing Techniques

New ways of designing products, systems, and goods are driving a small revolution in manufacturing that refers to the possibility to customize from the very beginning, not only the design, but also materials and machines.

The process of Advanced Manufacturing is driven by digital fabrication, which is a manufacturing process developed in an industrial context where precision and direct production is requested—for instance, automotive, industrial design and mechanical products—and that, nowadays, it becoming more and more frequent and adaptable in other sectors. It is connected to the change in digital technologies and their higher accessibility that enables a more direct production from design drawings thanks to parametric software and user-friendly production interfaces with flexible machines.

Along with software advancements, the use of computer numerically controlled machines (CNC) and robotic arms have opened new frontiers of investigation also in Architecture [4] within the field which can be defined as Mass Customization. Thanks to the bi-directional communication between digital information and material properties, design and fabrication tend to increasingly coincide, sharing common programming platforms and virtual environment.

Some of the actual instruments for Advanced Manufacturing includes CNC for milling and cutting, CNC for folding, CNC for moulding and Additive Manufacturing, on which we will focus with a specific case study in the next chapter.

Additive Manufacturing (AM) is a specific technique of incremental formation executed by the addition of material without using supplementary instruments in a process that is fundamentally opposite to the milling subtractive procedure.

This process offers a wide degree of flexibility and economic potential because the components are made directly from materials and can be therefore optimized in terms of the structure and quantity of material used. The main advantage comes from the possibility to produce unique components, which would not be economically sustainable to produce with traditional manufacturing techniques.

Currently, most of the 3D printing processes are based on plastic materials such as ABS, PLA, acrylate, photopolymer, polyamide (nylon), epoxy, polycarbonate and PMMA (acryl glass) or metal powder. Material mixtures might be modified for specific applications, in order to impart specific properties.

Polyamide, for example, can be modified so that it could be classified as 'in-combustible' and therefore used for aeroplanes. Some of these techniques have been later introduced to the general AM market making high performance plastics available, which can resist to high temperatures. The introduction of these novel plastic materials opens new potential applications for architecture and more specifically for building components of façades for example.

In construction, there is also a big ongoing research on powder materials like concrete, ceramics, clay, and organic materials that can be used to develop innovative products that can enlarge the possibility and tectonic of building systems in architecture.

Generally speaking on future developments of Advanced Manufacturing, for the last decade, the CNC production manufacturer has largely been an idle spectator to the advancements in AM systems while today, the idea to integrate the two types of production is becoming more and more fluid to make the hybrid production line efficient but flexible at the same time (i.e. Industry 4.0) (Fig. 3).

Fig. 3 The 3D printing of a cement-based full scale column has been tested to understand the potential of a Delta with a height of 4 m (ACTLAB, ABC Dept., Politecnico di Milano)

4 Optimized Building Components with Additive Manufacturing

The research at ABC Dept., Politecnico di Milano, ACTLAB (Architecture, Computation, Technology) research unit, fosters a performance driven approach where the material and production properties are incorporated in the design process.

Among new ways of manufacturing, Additive Manufacturing can really give the chance to match the economic viability of fordist and lean production together with a very high level of customization which can enrich a novel Mass Customization [8].

An example is an optimized structural component manufactured using Additive Manufacturing—and in particular Fused Deposition Modelling—which helps to acquire specific textural and geometric characteristics. A series of experiments have been conducted in order to establish optimal printing configurations with particular attention to the accuracy, mechanical strength, and optimization of production time. The idea is to design a building component systems which can became a wall, a façade, an envelope dry with variable shapes, inscribed in a working volume of more ore less 50 × 100 cm.

Each component is defined by microstructural patterns which have been developed on the basis of cellular solid lattice structure [5] and printed in PLA to simulate possible infill and volumetric patterns able to maximize mechanical behaviours while minimizing material consumption (Fig. 4).

The use of the Fused Deposition Modelling in the manufacturing technique ensures a good degree of precision measured with a tolerance of 0.1 mm. With this configuration, a single component can be produced using a medium-sized printer within a range of time that can go from three to max five hours depending on the speed of the extruder.

By basing the production on the use of different printers together, similar to a production line, integrated and coordinate, components can be produced and tagged in a pre-ordered series, ensuring easy assembly and structural strength during the whole process.

The design, development and implementation of the building components is a real example of Mass Customization based on a continuous interaction between the digital model, the finite element analysis and feedback of manufacturing machines. An innovative workflow and a systematic methodology have been also developed to inter-operate at different scales: morphological information, materials properties and performances are iteratively read, analysed and updated.

Computational Design and Advanced Manufacturing techniques has been used in this case study with the aim to 'inform' the design process from the very beginning with a reiterative process of design and fabrication which end up with an truly innovative workflow (Fig. 5).

Also the selection of the material plays an essential role in building technology choice. An important part of this research is the definition of optimal materials for the structural elements and in such a way as to bring the use of three-dimensional

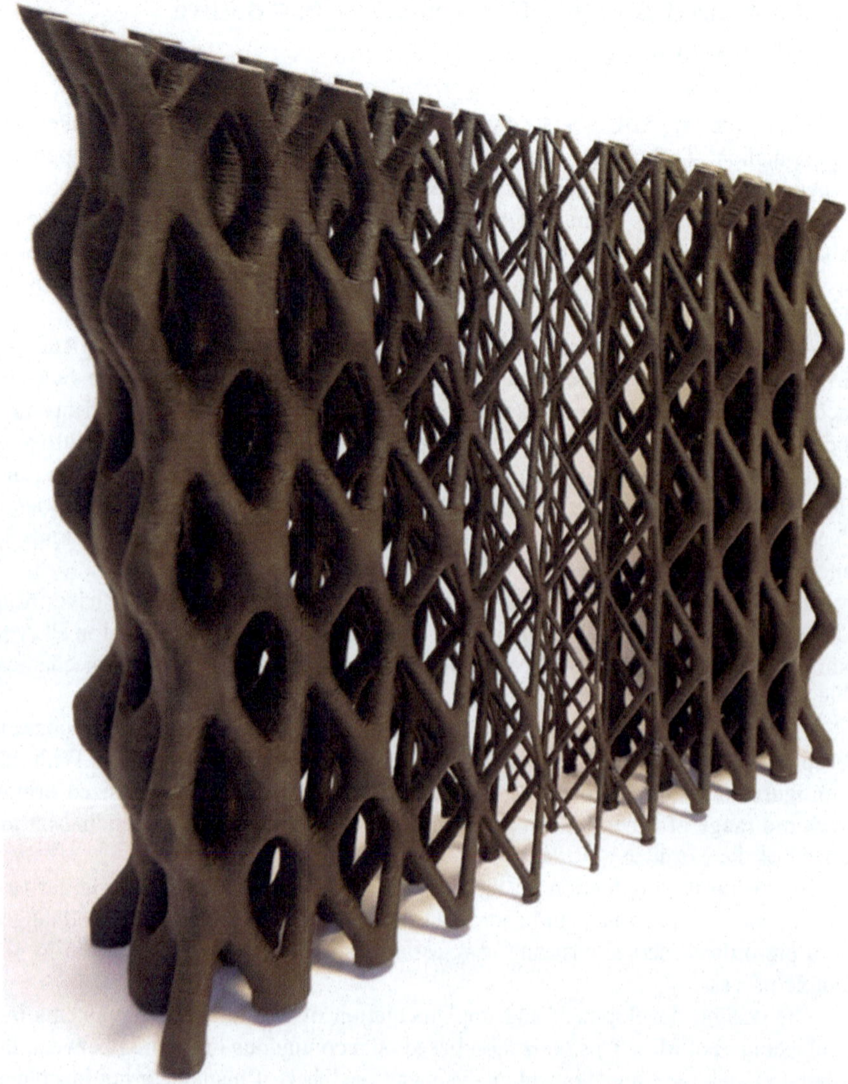

Fig. 4 Microstructural patterns based on cellular solid lattice structure have been developed and printed in PLA to simulate possible infill and volumetric patterns which are able to maximize mechanical behaviours while minimizing material consumptions

printing as a professional application. A key factor is a proper weight-strength ratio [7]. Several thermoplastic composites are currently being studied to evaluate properties and performance. A post-process of impregnation in epoxy resin has been tested to increase the rigidity and strength of the materials and then the moulded components (Figs. 6 and 7).

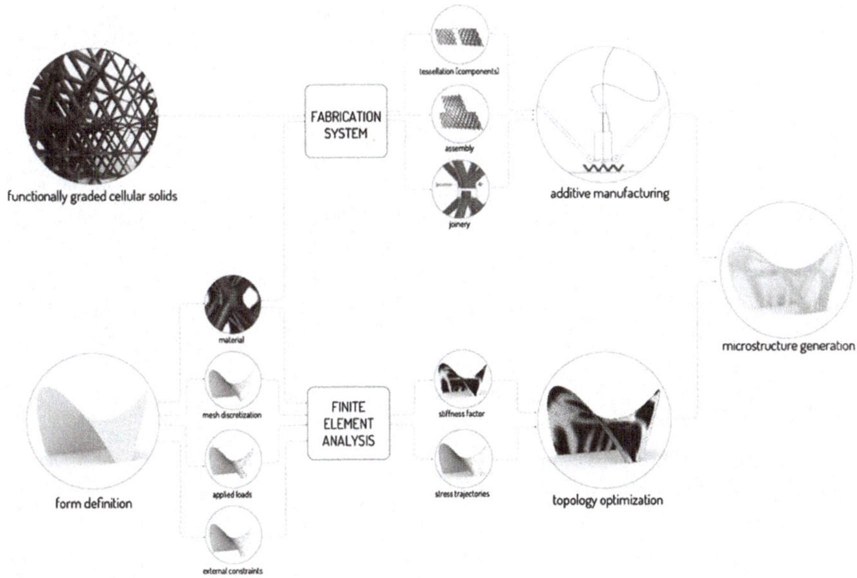

Fig. 5 The design, development and implementation of the building component is based on a continuous interaction between the digital model, finite element analysis, and feedback of machines for the realization of a pavilion at Politecnico di Milano (ACTLAB, ABC Dept., Politecnico di Milano)

Fig. 6 Material research, carried out by the composite thermoplastic tests with maximum performance, and post-production, using epoxy resin impregnation. (From *left*) Polylactic Acid, acrylonitrile styrene acrylate (ASA) and Polycarbonate (ACTLAB, Politecnico di Milano)

The goal is to develop a multi-performative construction system able to satisfy structural and environmental issues, integrated with its morphogenesis. The initial stage of the research focused on the development of a workflow to inform the infill microstructure of building components with meaningful load paths derived from topology optimization. Contemporary to this, several tests have been conducted over the study of potential infill insulation material and on performative tessellations as primary steps towards the development also of joinery system.

Fig. 7 Example of a possible application of the lattice structure for building a pavilion at Politecnico di Milano

From the analysis of the partial findings, it is evident that actual printing techniques present several constraints which should be properly analysed and modelled within the design environment in order to progress with the general goal of Mass Customization in Architecture and Construction.

5 Towards Immaterial Building Technology

Building Technology has always referred to a very traditional way of sharing information: this has been done through bi-dimensional drawings, excel files, paper information etc.

Today this workflow is completely obsolete, not only because ICT has gone very fast in improving tools and opportunities, but moreover because the request of information is nowadays faster, more precise, and long lasting. Information is not only used for different applications (structural, thermal, acoustic) but also to control the process (i.e. Building Information Modelling) and for long lasting maintenance.

In industry, this idea that information stays with the life cycle of a product is starting to be diffused. In Architecture and Construction it could be similar: what if we have information embedded inside a Building System to design it, produce it and to use it in its final placement in a building?

A sort of 'Immaterial Building Technology' is the forefront idea of the author and this paper, to underline that it is possible to include data in products of construction that starts firstly with the information of the material systems and its characterization, then to computational techniques to make it feasible, and then to fabrications constraints until service after delivery, in complete new and innovative workflows.

A building system or component could then be 'traced' from its design concept origin and could also give some information on its characteristics and perhaps also maintenance request in time.

This mean that computation, used to increase the level of calculation of performances, the quantity and typology of material, can also be used as a 'service' after construction, including a lot of 'immaterial' information in the materiality itself of the building system.

Building technology could then really foster innovation in the construction sector, not only thanks to increased performances but more thanks to the quantity of information it contains from its development that can be used a different levels in the construction chain.

This scenario that seems far is, in reality, quite near, as for the of design objects this is already possible thanks to IoT and will surely give a higher responsibility to designers, while increasing the knowledge of the whole AEC sector. We can conclude with a not so recent quote of *Richard Buckminster Fuller* which says, 'You never change things by fighting existing reality. To change something, build a new model that makes the existing model obsolete'.

Acknowledgements Thanks to Department ABC, Politecnico di Milano and ACTLAB Research Unit at Politecnico di Milano.

References

1. Ayres, P.: Persistent Modelling: Extending the Role of Architectural Representation. Routledge, Oxon (2012)
2. Borden, G.P., Meredith M.l.: Matter: Material Processes in Architectural Production. Routledge, New York (2012)
3. Fogliatto, F.S., Da Silveira, G.J.C., Borenstein, D.: The mass customization decade: un updated review of the literature. Int. J. Prod. Eco. **138**, 14–25 (2012)
4. Hauschild, M., Karzel, R.: Digital Processes (Detail Practice). Birkhäuser GmbH, Basel (2011)
5. Heinz, P., Herrmann, M., Sobek, W.: Functionally graded building Components. In: Production Procedures and Fields of Application for Functionally Graded Building Components in Construction. University of Stuttgart Institute for Lightweight Structures and Conceptual Design, Zukunft Bau, ILEK (2011)
6. Jiao, J., Zang, L., Shaligram, P.: Process platform planning for variety coordination from design to production in mass customization manufacturing, IEEE Trans. Eng. Manag. **54** (2007)
7. Naboni, R., Kunic, A., Paoletti, I.: Load-Responsive cellular envelopes with additive manufacturing. J. Facade Des. Eng. January (2017)
8. Naboni, R., Paoletti, I.: Advanced Customization in Architectural Design and Construction. Springer (2015)
9. Pine, B.J.: The New Frontier in Business Competition. Harvard Business Review Press (1992)
10. Shaub, K.: The Fourth Industrial Revolution. Davos Forum (2016)

Augmented Realities

Federico Alberto Brunetti

1 Introduction

The act of drawing, described as a convergent project term between arts and techniques in Renaissance humanism, has received a connotation not only of graphic expressiveness, but also of heuristic qualities of form, first of all in its mental conception and configuration.

The definition of *"inner drawing"* [*"disegno interiore"*] used by Federico Zuccaro[1] in the theoretical and doctrinal context of the cultural reformation debate of the sixteenth century images, identifies the consistency of an intellectual quality of visual form, a capable and necessary prelude in every concrete expression of visual and sculptural art.

In fact we could remember, following and redefining, as a conceptual metaphoric field of actual action following the ancient neoplatonic definition—between

[1]Federico Zuccaro (S. Angelo in Vado 1540—Ancona 1609). *Idea de' pittori, scultori et architetti,* 1607. See: John S. Hendrix, *Humanism and Disegno: Neoplatonism at the Accademia di San Luca in Rome*, Roger Williams University, 2007 ihendrix@risd.edu http://docs.rwu.edu/cgi/viewcontent.cgi?article=1000&context=saahp_fp.

See also: Federico Brunetti, *Del Visibile.* In: *Instructionum Fabricae et suppelleclis ecclesiasticae. Riedizione del trattato di Carlo Borromeo*, a cura di Massimo Edolo Marinelli e Stefano Della Tone, Edizioni Axtios Group—collana Monumenta Secondi Millenni; Roma, Libreria Editrice Vaticana, 2000 (pp. 405-416).

F.A. Brunetti (✉)
Politecnico di Milano, Milan, Italy
e-mail: brunetti@polimi.it

M. Hemmerling and L. Cocchiarella (eds.), *Informed Architecture*,
DOI 10.1007/978-3-319-53135-9_9

the two ancient antipode terms: "*disegno interno*" vs "*disegno externo*" (tr.: "*inner drawing*"/"*exterior draving*") as so called by Zuccaro (1539–1609). This binomial concerns the formal intuition that visual artists have about their work even before they express an outward representation, drawn or carved, that reveals in advance and come in action, the same final work. Therefore, the artist's work is recognized as founded not only on the reinterpretation of models, but rather on the creative intuition, a kind of visual background from which the work comes. In the same period when the path of consciousness towards the seeking of one's own identity and the research of a spiritual dimension were described as an "inner" site of an individual's qualities, it also becomes important for artists the idea that a sort of inner "vision" comes before the drawing, even of the heuristic one.

Although the aim of this writing is to deal with technological and contemporary matters, I simply want to remind metaphysic quality of the debate origins of the wide actual discussion about imagination and creativity in digital age and consider the importance to think about intangible and virtual characteristics of the new mode of the project representation. It is clear—through a series of steps which have gradually "automatized" the practice of the drawing: (both by reducing the gestural effort and facilitating the accuracy in handling geometric graphics tools)—that the drawing digitalization has speeded up, improved and interpolated the parameters of the process, enabling the dematerialization of the elaborate and helping multiple usability and visual empathy, with virtual and hyper-realistic features at the same time. The characteristics of algorithms—analogic or digital—that ensure biunivocal correspondence between drawing and reality—both as a relief and as a project [3] have allowed us to produce drawings of various types and for specific purposes (2D and 3D, plant sections, axonometric and perspective, photogrammetric surveys, and actually digital, photorealistic renderings and interactive animations, BIM models qualitatively integrated and selectively interpretable).

In addition, if the analog "executive" drawing was intended to be able to define all the structural details in order to run the project defined in its constructive realization—in the building site or in the factory according to appropriate standards of interpretative encodings—now, currently, digital drawings are integrating various procedures aimed at the automation of production (CAD-CAM, 3D printer, robotics etc.). Digital drawing shows potential for real process innovation—both in research and in technological/industrial applications—with respect to different aspects such as: accuracy in realization, execution speed, efficiency in risk situations, procedural continuity in different dimensional design scale (macro- midi- micro- nano-). Paradoxically, these exponential potentialities tend (permit and impose) to re-center the project issues around qualitative and ethical themes, rather than mere technological feasibility. In an era in which (theoretically) everything is possible, the more important questions are about what really deserves to be done and what concerns the human responsibility in the fields of social, ethical, and environmental consequences that each artefact implies.[2,3] In the present technological context, we can note that the drawing tends to be configured not only as a stored and observable artefact on an external physical support or media, but rather to be a representation in order to be accessible (ever since the preliminary concepts) as immersive experiences of a project in a virtual space embodied into the observer's perception and sensations. Hyperrealism in restitution and redundancy of information, however, is likely to lead to some difficulties in evaluating heuristic phases in the project results, not being able to observe the possible intermediate variants, or to understand and review the intermediate stages and the specific steps of the project that led to the final configuration. Particularly fascinating is also the high level of sensory simulation (in terms of picture quality even stereoscopic, like the acoustic integration, supplemented by implementary positional, postural and gestural interactivity.[4]

The person (designer or project commitment) is driven to live such an alternative and satisfying virtual experience where he could almost forget factual reality, with the risk to miss the complexity and unpredictability of real life, of objects and places of the world where we belong to.

2 From the Perspective to the Digital Interactive VR Frame

The history of perspective suggested the recurrent iconography of the visual representation apparatus in which the observer, through a semi-transparent framed screen, oriented towards the scene of real objects, is able to trace the shapes of images that appear—or graphically calculate—on the intermediate perspective screen surface.

[2]In fact there is a risk, apparently fictional or just theoretical, where you could set up an information omnidirectional circuit, interactive and globalized, between production, marketing and finance which will organize the production of artefacts in domains of the mere self-regulation (or rather its free transfer control distribution without any competent and ethical criteria) between supply and demand, so that it realizes a quite self-replicative methodology, as unfortunately several scenarios of devastating financial speculation arc showing in food and environmental terms. Sec for instance: Gael Giraud, *La Theorie des Jeux (Game Theory)*, Gamier-Flammarion, collection Champs-UT, 1st edition: September.

[3]2nd edition 2004. 3rd edition October 2009. Other works of the same author: www.gaelgiraud.net/en/.

[4]GPS (Global Positionig System), IPS (Indoor Positining System), Gesture and voice recognition etc.

2.1 Subjective—Objective, Perspective

The theory of projective geometry is the algorithm founding this process of graphic representation of the iconic description seemingly empirical methodology, characterized by an intense experiential similarity with vision. In this "optical-geometrical" device, the morphological quantitative and qualitative values of the object-described in line with the physiology of the eye (monocular)—interacts with the position and intentional view direction of the observer. The art historian Erwin Panofsky highlighted how deeply the perspective method is characterized by the distinctive correspondence between the observing subject, the object, and its representation in a mutual and irreversible feature of subjectification and objectification. Objectification is intended as the geometric fidelity by the precision of graphic rendering from a concrete data; and subjectification is conceived as the unilateral imposition of the point of view of the observer/artist space represented. This figurative power can be considered as an aspect of those changes carried out, and shifted as the cultural challenges of modernity, that Romano Guardini pointed out into the modern antinomy between service or domain.

2.2 Sensoriality, Body, Immersivity

In this interaction with the human senses, we can see how the natural bilateral symmetry of the person and the procedures for spatial orientation have been technologically arranged to reach the spectators for their reception in the most immediate and artificially spontaneous way. In fact, the spontaneity with which the person conducts its own existence is the result of a system of native interactions between the mind and the body. This real time experience in scientific terms is defined as: Proprioception "*The unconscious perception of movement and spatial orientation arising from stimuli within the body itself*" [9]. There are some gestures that best express this profound interaction between mind, gestural expressiveness, and visual feedback (perceived by all the senses in general, including spatial orientation and balance). The cycle between mental imagery, gestures, and real-time visual feedback allows a vast amount of spontaneous everyday actions, whose ability grows with the exercise of the experience, and also following the observation to other people at work as well. It is evident following the description of these simple and original manual activities and how the researches on artificial intelligence must be intertwined with those of neuroscience and learning sciences: in this range of Knowledge which tend to understand our behavior, explaining the natural and cultural learning mechanism processes at the foundation of our lives.

2.3 *Digital Visual Experiences*

The construction capability of virtual models elaborated by computer has been interfaced through the invention of the electronic monitor—and later digital screen—that allows to implement optical or graphic signals into visual information into the format of the devices frames, made interactive by pointer indicators.

The intensification of content of the digital technologies allows both innovative monitoring and new graphic processing by the representation of data in forms of analytical models of particular features: instant three-dimensional graphics (dynamic sequences), qualitative interpretations (color thematized), adjustable spatially (perspective views). Furthermore, the development of digital display.

Systems, not only on the 2D or 3D on computers flat screen visualization but even through stereoscopic immersive and augmented reality technologies, are demonstrating to be a tool not only for a better understanding of the complex interweaving of the morphology of the design in computational space, but also a way to communicate the methods of visual innovation borne in the interweaving of visual arts and science. The computational innovations of algorithms allow a better understanding of the shapes of design inventions or even science invisible data events and may themselves become a form of expression rationality and creativity in a new way of being able to see.

2.4 *Virtual Vision*

Upon this concept, it is even more amplified when duplicated by the stereoscopic effect of shifted double perceptive vision and motion sensors that interact with the body movements into the simulated virtual scene, based on the digital technologies defined as immersive vision. This means the possibility to see an event that could be optically shown to the observer, as projected from a frontal opaque screen, for a virtual and interactive perception, but essentially excluded from the surrounding environment. In particular, there are two main types of viewers available, some based and compatible with the Android operating system, referred to the so-called "VR" (Virtual Reality) and "AR" (Augmented Reality). Both offer undoubtedly remarkable communicative potential. These visual digital immersive stereoscopic experiences, previously realized into interactive "stereo-rooms", are nowadays included in the design shape of digital visors with binocular stereo screens, with accelerometer and gyroscopic panoramic view, so related to the user's head and body for environment simulation. These immersive vision digital technologies are defined also "virtual reality", and with the acronym "VR", are based upon this concept—even providing stereoscopic depth effect—interacting with motion sensors with the observer body movements into the simulated virtual scenes.

2.5 Augmented Reality Vision

Recent typologies of digital stereoscopic viewers have been realized—especially designed for observing 3D digital interactive content images projected through a system of transparent optical prisms—which allows the simultaneous perception of "Augmented Reality" (AR), where images and lists of information are overlapped by see-through glasses on the immediate vision of the surrounding environment; in these see-through visual devices the digital screen become as a transparent window.

These devices are interlaced with a series of built-in sensors for positioning and movement, 2D (and furtherly 3D) cameras for pattern and shape recognition or code configurations, and suitable for gesture and vocal commands. Such scenery of integrated visual knowledge must be supported by graphic interfaces and appropriately designed app configurations, to allow guided experiences or inquiries, pre-engaging both connections on remote and optical links on-site.

These hypotheses of interpreted overlapped visibility must be of course supported by suitable graphic interfaces and appropriately designed apps configurations to allow guided experiences or inquiries, prearranging both connections and on remote rather than optical links on-site, or even shape recognition sw. with unprecedented effectiveness results. The whole tangible reality could potentially become so an "index" that could be recognized and traced to integrate virtual content, contextualized with the effective environment. In this direction, following procedures already well tested, it is possible to interact with digital positioning codes placed into the existing real environment, and making the observer able, by index recognition technologies through GPS, IPS, or QRC (Global Position System; Indoor Positioning System; Quick Recognition Code and barcodes), to connect, receive—and "see" overlapped—specific targets information by the implementation towards databases of images, even in 3D stereo, or other documentary sources or digital contents. It is methodologically relevant also to remember that the vision of these see-through glasses is not realized on an opaque screen (as in VR), that other devices could blind the overlooking space vision, but ingeniously projected—side-by-side—by two miniaturized projectors directly to the observer's retina. Picture shooting skills are integrated into the micro camera in the glasses frame, which is also designed for transmission even in conference call by WiFi, or photo/video recording into internal memory.

Recent researches are oriented, for example, towards the shape recognition, by means of the Hough transform, in order to engage them as identifying sources-from digital captured images—into mathematical algorithms for augmented reality applications (Figs. 1 and 2).

Fig. 1 (*left*) Augmented Reality: a see-trough smart-glass device can recognize optical codes by its integrated camera and connect the Wi-Fi reachable related information overlapping them on screen to the direct view of the scenery. The visual experience can be shared with other partners, both in the same environment, rather than in remote connection

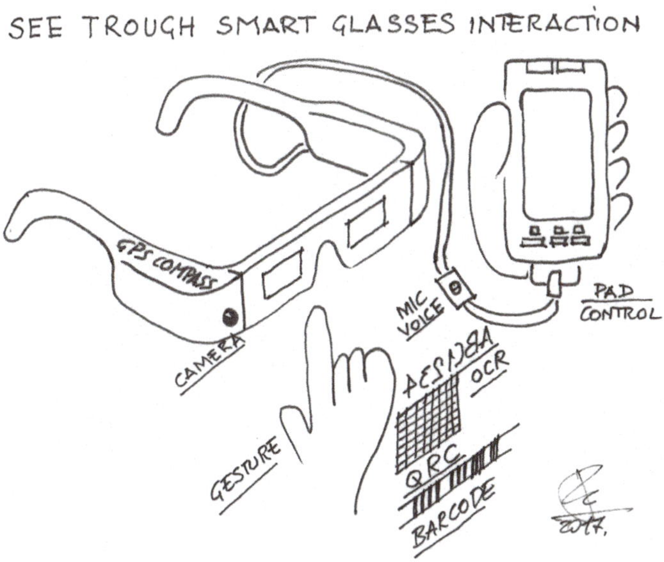

Fig. 2 (*right*): Smart-glasses view and camera assisted environment information. Interactive visual communication smart glasses experience: GPS—IPS, voice or gesture command, pad alphanumeric typing

3 Embedded Vision

This new compound terminology defines a new area of hardware and software development researches oriented to integrated technologies able to share the results of image recognition, supporting on decision making and other sort of integrated sensors, for autonomous applications. In this context, the human presence is delegated to a supervisor role, related to the automated ability to explore, and be potentially self-organized, by the machine. "*Embedded Vision refers to the practical use of computer vision in machines that understand their environment through visual means. Computer vision is the use of digital processing and intelligent algorithms to interpret meaning from images or video. Computer vision has mainly been a field of academic research over the past several decades. Today, however, a major transformation is underway. Due to the emergence of very powerful, low cost, and energy-efficient processors, it has become possible to incorporate practical computer vision capabilities into embedded systems, mobile devices, PCs, and the cloud. Over the next few years, there will be a rapid proliferation of embedded vision technology into many kinds of systems*".[5]

4 The Visual Experience Between Art and Science

Coming back to the starting point of this comparative discussion, we can evidence how the same original imaginative concept is still deeply actual, even if implemented in different technological and conceptual interpretations. A characteristic of the contemporary audio-visual technologies, supported by the versatility of multimedia digital platforms, it is certainly an attempt to wrap the spectators with the most likely scenery: multi-sensory, and possibly interactive by gestures, or even tactile, even though not yet easily olfactory.

This tendency to the immersive ability has coincided with a progressive alienation and exclusion of the spectator with respect the surrounding reality (itself however an archetypal condition of the theatrical scene or the same darkness into the movie theatre). Furthermore, these features have generated—particularly in the devices of the early generation of presentations—virtual reality experience and objects tending to progressive isolation and alienation: helmets, goggles, gloves, cabins: objects that are particularly attractive but also disquieting (Fig. 3).

[5]http://www.embedded-vision.com.

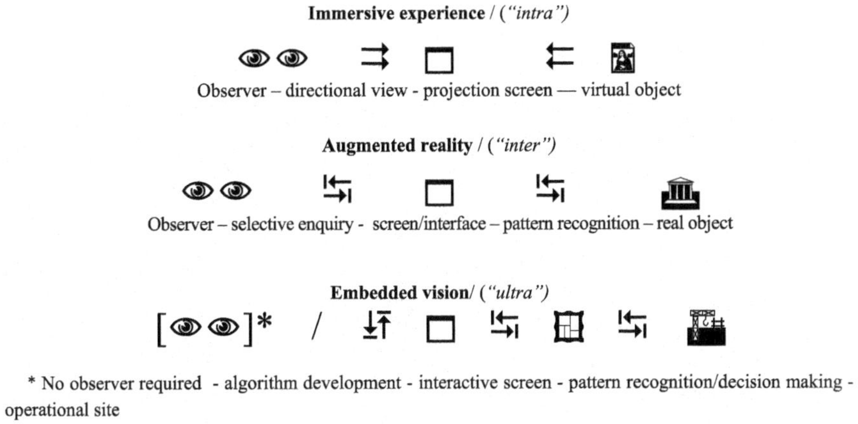

Fig. 3 Concept diagram: the frame— as a perspective and computable outline— at the *fulcrum* of contemporary vision algorithm. Of course this notation is intensified in this and simply means that the devices of this genre are progressively autonomous in their own activity: the algorithm is obviously generated by a human developer and the integration between software and hardware, even in more and more miniaturized, portable or even "user wearable". These will be a real design challenge for new form of service to the people needs

5 Image and Representation

Once these original assumptions are clarified, it can be evaluated, with as much awareness as possible, the strategy to exploit the opportunities of new technologies, such as those upcoming 3D visual devices, that enable interaction visual, audio, gestural and proxemics and even with the surrounding environment.

AR glasses could allow us to observe the surrounding reality with an "overlap" in 3D vision of another scene or object designed seem to be able to become the ideal tools for the vision of designers, accustomed with "the compass in the eye" to feel the proportions of the composition in the space, with the required harmonies and balances almost even prior to drawing on white paper, just like the "watch over" as a mental attitude of the artist, as likened to perhaps the scientist, is a way of looking at the present, investigating every possible information, linking the data of experience with those of possibilities, those of the visible and the invisible, of the presences and absence, of the evidences and clues.

Yet we must not forget—in terms of history of material and technological culture—the original genealogy where the 'glasses' belong Thomas Maldonado highlights. Thanks to the application for the purpose of the eye lens, the scientific biography of the intellectuals of the late Middle Ages has been transformed and prolonged significantly, in comparison to the average of the operability, thanks to the new ability to read and write as writers and scholars in mature age.

Without necessarily evoking the European medieval concept in which the view was seen as a sense ultimately liable to deception—where rather the sense of touch as the sense organ was considered more "tangibly sure"—it can be considered as the attempt to approach to visual perception "by wrapping it" by more and more credible and plausible representations in relation with the direct experience, ultimately groped to reach the core of the imagination in which every vision is collected and from which every prefiguration is generated.

The interaction between vision and stereoscopic vision in motion, from stereo and assisted by the possibilities of interaction with camera and control WiFi, could come close to perceptions well useful for the formation of the mental imagery that our intelligence elaborates to empathize with and become participant in the shapes of the external world.

Similarly, it is from this deep archive of imaginative activities that the ideas of the invention, emerge selectively, or any prefiguration of project and innovation of design. Also worth mentioning is the possibility, given the predisposition of WiFi, to be able to create a network to share these experiences between different stakeholders in the mutual interaction.

Far from any naive and enthusiastic drift towards so-called augmented reality (AR), I would argue that we should consider with sympathetic attention—and with the awareness of a minimal degree of historical detachment—about the fact that in every age of the history of images has been deeply interweaved with the modality of vision of its own period purposing discreetly and far-sightedly to observers who know "how to look". What new and foreseeable digital divides await us in this scenario of the de-materialisation of knowledge? Will we see more and better or too much?

Since the human vision experience is natively an interaction between the optical sensitivity (and recognition through memory) of places, people, objects, jobs (and other infinite things), in the same way, the so-called "Augmented Reality" (AR) could became the field of deeper relations between the person and the disseminated knowledge existing in the web. If appropriately implemented, these relations can become a very spontaneous way to explore the meaning of things, and valorise their best practice usability.

As how in the late middle-ages, where glass craftsmen obtained through the dissemination of glasses and higher longevity of scholar activity which later supported Galileo's invention of the telescope (and the consequent new way to represent the entire world), in the same way, our applications of (so-called) Augmented Reality can be hopefully the first platform for new kind of unimaginable relations and jobs for people. Not any ingenious hope, but a clear desire and responsibility for visual researchers.[6]

[6]Federico Brunetti, Facoltà del Design Politecnico di Milano, INDACO Department. *Excursus in Aveiro. Digressão em Aveiro. Arte-Design-DESEGNO-Técnica-Ciencia*. Workshop of Drawing and a theory seminar about the relations between Drawing, Arts and Science. 16.10.2006, University of Aveiro, Campus Universitário de Santiago Aveiro (Portugal) Department of Communication and Art.

6 Drawing Didactic: The Sight to the Real: Theoretical References and Working Methodologies

To introduce some concepts about the didactic experiences and protocols about how to share these competences to the next generation's technologies users and actual students of our Schools, I need to introduce and describe some of the concepts I regarded as theoretical criteria on which to found the thoughts, the work methods, the interdisciplinary comparisons (correlations) and the working applications in the teaching of drawing.

I would also like to put forward a sort of metaphor bearing in mind that there exists a significant analogy between activities and abilities relating to the act of writing and that of drawing.

6.1 Read-Write/Observe-Draw

The semantic content of verbal thought can be mnemonically preserved and transmitted through space and time by means of the written support. Similarly, visual thought can convey perceptive and imaginative experiences connected with visual and spatial forms by the act of drawing. Thus, when one learns the structures of the textual elements, the ability to read and write becomes possible, and in time there seems to be a spontaneous immediacy in the relationship between thinking a speech, being able to read it, and knowing how to write it.

Similarly, it is possible to build the abilities to translate one's observations of the real world—or the prefiguration of invented forms—into drawings, by learning the logical-formal structures of geometry, as well as free-hand drawing, or the perceptive effects attainable through rendering techniques, thus becoming able users of that universe of signs which is the object of our discipline.

6.2 Description—Interpretation—Prefiguration

In this context, looking at the use of the graphic language as a means of cognitive representation, it could be interesting to distinguish certain levels of increasing complexity and abstraction in the relationship between the thinking individual and the surrounding world, which correspond to similar level of in-depth analysis. In contemporary epistemology, within the trinomial: description—interpretation—prefiguration, it is accepted practice to identify the three criteria of increasing involvement between the subject who reproduces the object (through drawing or other means of representation) and the object (that already exists or will be projected) of such representation.

Description can be defined as the transcription of the most extensive collection of data the "observing" subject is able to capture/extract from the object being observed in the real world, through the expressive characteristics of the languages and the technical instruments at his disposal. Interpretation must be understood as the critical selection of the data collected, separated, and actively directed/applied by the subject according to his working needs.

Prefiguration is the invention of new possible configurations of reality starting from the collected data, which are tampered with, modified, or related to each other according to a specific project, obtained by sublimating the state of things and the pre-existing real world to achieve the project scenery.

Clearly these three cognitive criteria must be considered as a hypothetical procedural scheme, replete with possible interferences and retro-actions, since it is often necessary and suitable that these three actions take place in a mutually synergic and communicating fashion. In fact, the description cannot be naively neutral, but it carries within the instrumental methodologies and the thematic approaches characterized and conformed by the disciplines and the languages used.

Archaeologists have a saying that describes the invasive exploration and excavation procedures of research sites: "you find what you are looking for"; or one can discern what one can recognize.

Cognitive competences and descriptive abilities are inescapably intertwined. Similarly, the selective modalities of interpretation must inevitably conform to the project's goals as they mature, thus creating in the mind of the observer-planner destination, which makes some elements more interesting than others, so as to stress the need to obtain further information on portions of reality or on different and heterogeneous orders of magnitude.

Finally, by continuity or by opposition, the acts of prefiguration—which methodologically consist of the act of overcoming the existing conditions—must be based on the palpable and local context of actual, expressed, or latent relationships and needs to which they answer.

More precisely, the problem of invention finds its place in our discipline as a factor that characterizes the intelligent forms of life, and therefore originates the projecting ability. Such ability can be placed at the intersection of some pre-conditions, summarized as follows:

- the memory of a logical-formal archive of pre-existing elements,
- the ability to experiment new relations among elements which up to that point were not deemed mutually pertinent,
- the ability to intentionally or fortuitously uncover new factors in the fields previously explored.

6.3 Drawing Concept and Taxonomy

It seems interesting to propose here a conceptual map that has already been made more than a decade ago (a long time in the light of technological developments) that I had developed for a possible taxonomy thematically classifying all texts relating to the design in the libraries of the Politecnico di Milano, following their centralized digital cataloguing. The three main poles possessed, in a conceptual space, the research directions that every text indicated, and reference intermediate disciplines of such thematic orientations. Such a ternary pole arrangement seems to me to prove still methodologically valid even for what might be called as the virtual design experiences, conceived as a central cross science, and indeed it can posed as an interesting reference for understanding the variegated technologies and application functions, of devices and software, that are proliferating: with differences sometimes essential, and often instead with obvious recurring analogies or genealogies, despite the diversified commercial garment with which they are presented.

Conceptual Areas. Drawing as Conjectural Practice
(Federico Brunetti 2003)

Perception – Communication
Symbolic representation
Neurosciences
Artificial Intelligence

ALGORYTHM
Geometry
Science / Technology
Analogical / Digital

THEORY
Philosophy
Aesthetics
Word /Text

↖ ↗

DRAWING

↓

Philosophy of Science History of Art
ERMENEUTICS
History
Archives
Monument / Document

Conceptual Areas. Virtual Drawing experience as a Design cross-science
(Federico Brunetti 2016)

ALGORYTHM
Software developments
Immersive technologies / virtual prototyping
Materic behavior simulation

THEORY
scope / usefulness / sustenibility
global icons visual meaning
Big-data resources / social forecast

↖ ↗

Virtual DRAWING experience

↓

ERMENEUTICS
Archives / databases
Document / Monument
exinting nurbs databases / virtual prototypes catalogues

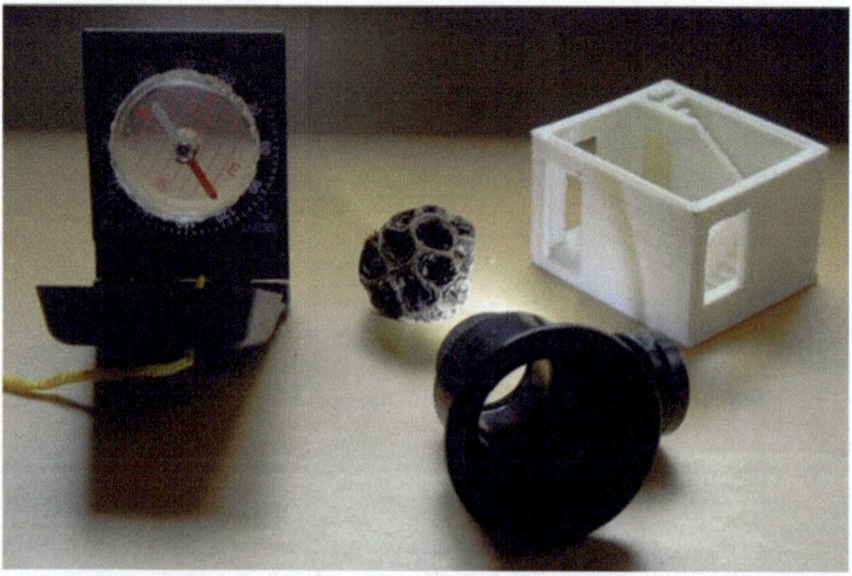

Fig. 4 New directions of project drawing under the enlargement lens of immersive technologies: Geo Positioning, organic morphologies, prototyping 3D print preview

6.4 Open Ending Consideration. "not Done by Human Hands."

"Not done by human hands". This definition, attributed to some sacred icons referred to the Face of Divinity in order to introduce a mystical contemplation factor or transcendence from any human imitation of reality, takes an unusual actuality when applied with a literal transposition to the drawings or objects that are no longer being artefacts or manufactured articles, but processed—more or less independently by the machines. In this era of nanotechnology, we know how some nanomaterials may be able to self-replicate, picking up from the environment the material elements needed to reproduce indefinitely equal parts of them; these self-referential dynamics of such scenarios—including robotics and materials science—obviously offer new concerns and obviously even ecological matters of discussion about the design power of these technologies (Fig. 4).

Without further ado in these issues, however far beyond the reach of our discussion matters, we see something unusual in the objects presented in the educational collection CITNB.[7]

---

[7]Exhibition *Computation is the new Black. Updating the Making of Architecture, organized by Prof. Marco Hemmerling* at the School of Architecture Urban Planning Construction Engineering (AUIC) of the Politecnico di Milano in January 2016.

In fact, as far as programmed by skilled operators and designers able to govern the engineering process—and in relation to specific materials and contexts—these objects are offered as example of potentially self-generated by an internal algorithmic rule, much more than just obtained due to some type of manufacturing ability. Even without delaying in the rhetoric tayloristic antinomy between craft and machine, their specific qualitative value lies in following a rule, or internal program —albeit minimally sensitive to the forces of the field in which they are—as it could also be said of nanometric regularity of a mineral crystal, or spirals in a fossils of leaves of a fern or hull of a seed, as a diatom. For these reasons, the precision and reiterated shapes of these didactic models of projects goes far beyond the invention or algorithmic acquisition of some harmonic canon, but aim to discreetly approach the nature itself, participating fairly in the re-creative work, otherwise called creativity.

References

1. Barocchi, P. (a cura di), Paleotti, G., Dal.: *Discorso intorno alle imagini sacre e profane*, Bologna 1582, libro 1, In: *Scritti d'arte del Cinquecento*, Milano—Napoli (1971)
2. Vitta, M.: Il sistema delle immagini: estetica della rappresentazione quotidiana. Napoli (1999)
3. Brusatin, M.: Disegno/Progetto. In: Enciclopedia Einaudi, vol. IV, Einaudi, Torino (1978)
4. Brunetti, F.: Through the screen. Concepts and typologies of digital vision experience. In: Proceedings of the 15th Conference APLIMAT (2016)
5. Anceschi, G.: Monogrammi e figure: teorie e storie della progettazione di artefatti comunicativi. Firenze (1988)
6. Martin, K.: The science of art. Optical themes in western art from Brunelleschi to Seurat. New Haven and London
7. Panofsky, E.: Die Perspektive als "symbolische Form", in "Vorträge der Bibliothek Warburg". Teubner, Leipzig-Berlin (1927)
8. Guardini, R.: Das Ende der Neuzeit: ein Versuch zur Orientierung, Würzburg, Werkbund-Verl., 1965.; tr. it: La fine dell'epoca moderna. Il potere, Brescia (1950)
9. Dickinson, J.: Proprioceptive control of human movement. Princeton Book Co. (1976)
10. Brunetti, F.: Experience of three-dimensional vision in the era of digital interaction. new devices: opportunity and challenges for applied mathematics and visual design. In: Proceedings of the 13th APLIMAT Conference (2014)
11. Brunetti, F.: LHC between science and architecture. Milan (2009)
12. Zhang, Q., Tan, H. S., Luo Y.: "Imaging of moving target with rotating parts based on the Hough Transform". IEEE vol. 46, No.1, (2008)
13. Ferraris, M.: Dove sei?. Ontologia del telefonino, Milano (2011)
14. Maldonado, T.: Gli occhiali presi sul serio, (contributo al ciclo di conferenze Sapere e narrare). L'uomo e le macchine, Firenze (1999)
15. Merleau-Ponty, M.: Phénoménologie de la perception, Paris (1945); Le visible et l'invisible, Paris 1964. L 'OEil et I 'Esprit, Paris (1964)
16. Mitchell, W.J.: The Reconfigured Eye, Visual Truth in the Post-Photographic Era. MIT Press, Cambridge (1994)
17. Dorfles, G.: Horror pleni. La (in)civiltà del rumore. Roma (2008)
18. Brunetti, F.: Attraverso il Disegno. Relazioni tematiche, tracce metodologiche e atti didattici per l'architettura ed il design. Reprint with translations, Milano (2012)

19. Brunetti, F.: "Attraverso il Disegno", presentazione di un itinerario didattico. Esperienze per la formazione al Disegno. In: Proceedings of the 10th APLIMAT Conference, Bratsilava (2011)
20. Bailey, K.: Typologies and Taxonomies—An Introduction to Classification Techniques. Sage University Papers: Sage Publications, Thousand Oaks (1994)
21. Haeckel, E.: Kunstformen der Natur, "Artforms of Nature" (1904), Freedom in Science and Teaching (1879), The History of Creation (1868)

Computation in Architecture: Potential and Challenges for Research and Education

Jens Böke

1 Introduction

Modern computer technology has a decisive influence on the building and construction industry. A vast range of digital tools are nowadays available to all disciplines involved in building and construction processes. This impacts not only the planning and manufacturing processes but also the quality and the appearance of the results that *can* be achieved. Examples include the complex geometries of designs and the simulation of requirements concerning statics and building physics or, also relevant, the organization and communication as part of the project. Designs can today become directly and precisely manufactured via digital fabrication. In this context, buildings and constructions are promptly realized which would not have been feasible without state-of-the-art computational tools. The last decades have been typically characterized by an architectural and design language borne from the capabilities of this technology.

The actual added-value created, by the application of computer-based tools in architecture, is the topic of an ongoing controversial discussion. Only a few decades have passed since they were introduced in the course of the Digital Revolution. Compared to the wide range of new methods and possibilities, we have very limited experience as to the efficient and sensible use of these technologies. In the past, the focus was mainly on the digital tools themselves. Studying them resulted in new developed strategies which were then transferred to exemplary implementations. The future relevance and acceptance of digital strategies will depend on an obvious added-value for architecture. Next to architectural-aesthetic possibilities of geometries or complex structures, an added-value might lie in an increase of design

J. Böke (✉)
Technical University Delft, Delft, The Netherlands
e-mail: jens.boeke@th-koeln.de

M. Hemmerling and L. Cocchiarella (eds.), *Informed Architecture*,
DOI 10.1007/978-3-319-53135-9_10

and in construction process, efficiency as well as a higher quality of achievable results. In this context "performance" has become an established term. In order to increase the performance, it is necessary to focus considerations not on the capabilities of computer tools, but on the requirements of the building project.

Digital tools constitute a new challenge for architects, planners, and engineers. In addition to the traditional proficient knowledge in their fields of expertise, professional knowledge of information technology (IT) is increasingly required. Classical career profiles like those of architects might have to be reconsidered with this background and new disciplines might become necessary such as expert planners to become the interface between architects and IT, involved in the planning and construction processes of buildings. Today we face the third digital revolution, a comprehensive digitalization and intensive networking of our environment. The current debate is driven by concepts like: "Internet of Things", "Industry 4.0", and "Big- or Smart Data". It also affects architecture since, in the front of an increasing building automation, the influence of computer technology goes beyond the planning and construction phase and now also has an impact on the building's operating phase. Computers make "intelligent" control of essential functions possible and often represent the core of "smart architecture". This however results in new responsibilities for architects. Traditionally, the architect supervises the design phase and the construction processes of a building up to the handing over to the owner. Automation now means new tasks during the utilization phase of the building. Within the matching of intelligent automated building technology with the building's usage, and, in the definition of control parameters according to user requirements, there lies a new responsibility [1–5].

2 Research and Teaching

Education is particularly important. Appropriate teaching equips the professionals involved in the building processes with the necessary IT expertise. This is also relevant since, preceding Building Information Modelling (BIM), the computer has taken up a central position as a communication and working platform. The new requirements become recognized nowadays and some universities are already offering study courses in architecture with a focus on digital tools. An example is the post-graduate program: "Master-Computational Design and Construction" (M-CDC) at the East-Westphalia University of Applied Sciences in Detmold. Not only is the mediation of existing knowledge important, nonetheless, the rapid introduction of computer technology into architecture brings up many new open questions. This is where research plays a key role. The better we understand both the chances and risks of the technology, the better we can develop uses that will lead to added-value in architecture.

3 Research ThinkingSkins

Implementations of digital controls for adaptive building envelopes are investigated in the research under the title "ThinkingSkins". The façade has a substantial influence on the appearance of a building. Beyond that, it also plays a key role in the energy performance and the user comfort of a building as it mediates between outer and inner spaces [6]. This means coping with constantly changing external influences and interior requirements. The dynamic adaptability of the building skin increases its efficient performance. Today, external influences can be measured with sensor technology which triggers appropriate digital reactions. They are called appropriately: "responsive facades". Computer technology allows for the control of individual, reactive functions and their interconnection. The matching of the interactive reactions with the use of the building represents a challenge for the design of the façade. The study of ThinkingSkins is based on the assumption that strategies of digital, rule-based design can contribute towards an effective total system of individual reactions. The aim of the research is a higher performance of building envelopes reached by the design ability of self-regulating complex systems that take indoor requirements into account [7].

A special potential lies within the linking of research and teaching. Therefore, some studies in the research of ThinkingSkins are carried out in collaboration with students. Several workshops were held in this context on the topic of responsive Building envelopes. An example is the workshop "façadetronics". It was part of the efnMobile program, a workshop series by the European Façade Network and was held in November 2015 at the East-Westphalia University of Applied Sciences in Detmold. The workshop was inspired by the conference "facade2015—Computational Optimization" and was part of the local façade week. Students of the postgraduate Master courses International Façade Design & Construction (IFDC) and Master-Computational Design and Construction (M-CDC), as well as students of the façade Masters at University of Luzern attended the workshop. The aim of the program was the development of innovative automation concepts for the facade. These should also, in addition to a theoretical examination, be implemented and studied in the form of prototypes. For carrying out the task, international and interdisciplinary teams were formed (Figs. 1, 2, 3, 4, 5, 6 and 7).

The starting point of the investigation was not based on the options of available technology, but on the existing functions of the building envelope. Therefore, each team got a façade function to edit which was assigned by lottery. The selection of functions was based on a model developed by Klein [8], the façade function tree. For a focus, the selection of functions was limited to those in the context of the influencing factor, that is the sun, such as glare protection or sun shading. At first, the teams examined how the corresponding facade function has been traditionally implemented. This included both, the analysis of existing designs, as well as the determination of physical principles. On this basis, the first ideas arose on how the

Fig. 1 Workshop "façadetronics"

Fig. 2 Team discussion as part of the design process

informing and activating of facade structures may contribute to the fulfillment of the respective function. The identification of relevant information and the appropriate sensor technology was as relevant as the development of active components with built-in actuator technology. It was also part of the task to develop a concept for the control strategy. It had to be decided; how gathered information is to be transferred

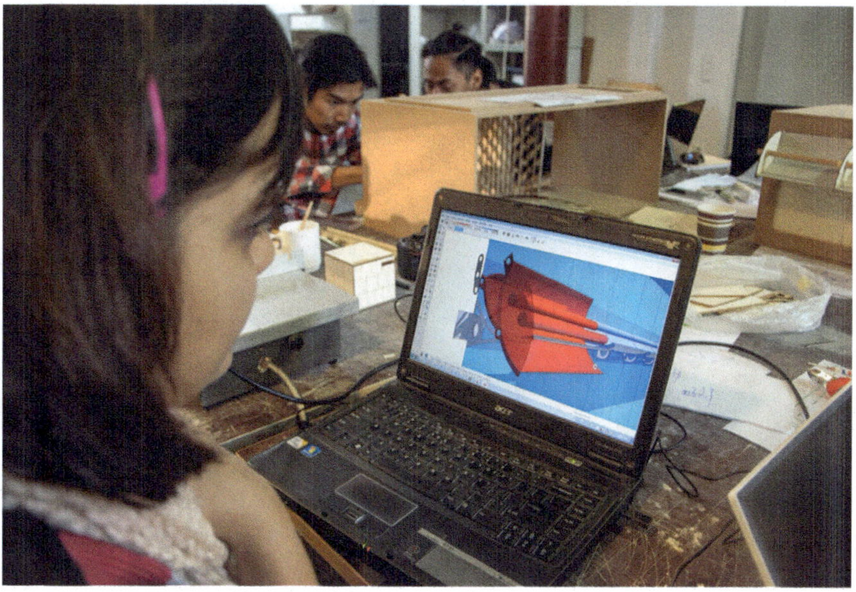

Fig. 3 Digital design of the structure

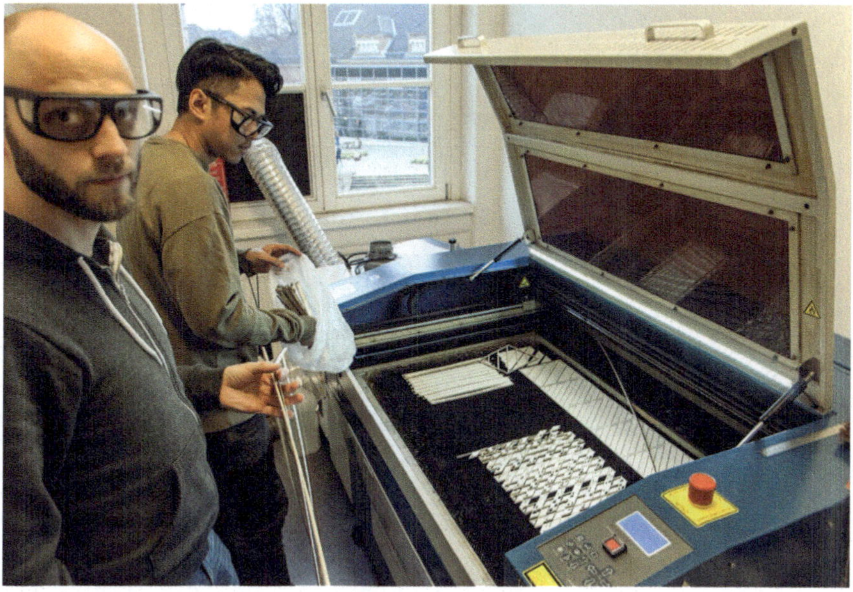

Fig. 4 Realization of complex components by digital fabrication

Fig. 5 Prototyping

Fig. 6 Implementation of the Arduino-control

Fig. 7 Project DOUBLE LOUVRE prototype

to a corresponding reaction of the facade. The challenge for the students was to develop both, a concept for the physical construction as well as for its digital control. On the physical level, the teams were faced with the merging of kinetic and electronic components. Digital design strategies were applied and led to 3D-models of the structures. The subsequent digital fabrication enabled the realization of complex components. The concepts for the digital controls were first graphically developed and visualized by dependency graphs. A later implementation of the control in the prototype was based on the Arduino platform.

As a result, any concept was graphically displayed on three posters. The first poster includes the analytical examination of the façade function. The second poster documents the concept and its principle of construction. The visualization of the design is based on the 3D-models created earlier. The third poster illustrates the development and installation of the prototype. The prototype itself is an important part of the design result. The implementation of the prototype took place within a prefabricated model framework. Through this, the students were able to concentrate on the construction of their active structure. By specifying the prototype size and shape comparable and combinable results were obtained. The model-scale prototypes allow the physical visualization and verification of the drafts functional principles.

For example, a project within the workshop led to the idea of a "PARACHUTE FACADE". The aim of this construction is a self-regulated shading. The design of the students Amir Saadatfard (M-CDC) and Bahman Bidmeshki (IFDC) proposed a grid of rectangular frames that is equipped with stretchable fabrics. A stepper motor is positioned at the center of each frame, supported by the fabric itself. It opens or closes the textile according to the daylight situation. The intensity of sunlight is measured per item, based on the information gathered by photo-resistors. The information becomes transformed into the reaction of the stepper motor by a microcontroller. By embedding sensors, actuators, and computer control, each frame performs as an intelligent device that can respond to a changing intensity and position of the sun. The project "DOUBLE LOUVRE" by Ivan Cakaric (IFDC) and Nathanael doubt (HSLU) deals with the theme of light control. Their design contains two concerted reflectors that are mounted on the facade to optimize the natural illumination of offices. The system is matched digitally on the daily path of the sun. Each reflector has a non-reflective back, with a protective function in case of excessive sun exposure. Depending on the intensity of the sun, the reflectors change their positions to direct extra natural light into the interior when needed. The embedded digital controller navigates the successive tuning of the reflectors in view of the current daylight situation.

Seven more results emerged within the workshop, in which the potential of informing and activation of cladding components were investigated. The results were presented to a professional jury and exhibited within the conference "Facade2015 – COMPUTATIONAL OPTIMISATION" [9, 10] (Fig. 8).

Fig. 8 Result of the project sliding shield by *Julkifli Axel Siswoyo and Vanvithit Bhukanchana*

4 Conclusion

Computer technology offers many new opportunities in architecture. In order to meet the challenge of gaining most benefit from these possibilities, it is essential that experts become qualified, moreover, strategies for the technologies application become researched. Today, new tasks and new responsibilities derive from the extensive digitalization. They are also met in the research of ThinkingSkins by the transfer of digital design strategies to the automation concept for the building's usage phase. The confrontation of students with current research issues lies a great potential for both sides. Students are aware earlier of current issues and future requirements in their field of study while they generate knowledge by the experimental use of new technologies. It is assumed that not every concept of the workshop "façadetronics" can be converted into performance-enhancing components and design principles for façades. The projects show, however, due to available and cost-effective technologies in the field of information technology and electrical engineering, new potentials and possibilities arise to actively cope with the dynamic influences on the façade. The workshops show the extent of the required expertise, necessary for the development of active systems. The need for skills in the development of components and constructions, electrical and control engineering, as well as in information technology and programming shows the necessity of interdisciplinary cooperation in the construction industry. Against this background, the ThinkingSkins workshops contribute as a scenario for multidisciplinary collaboration by bringing together students from different disciplines.

References

1. Hemmerling, M., Tiggemann, A.: Digitales Entwerfen: Computer Aided Design in Architektur und Innenarchitektur. Fink, Paderborn (2010)
2. Hauschild, M., Karzel, R.: Digitale Prozesse: Planung, Gestaltung, Fertigung, 1. Aufl ed. Inst. für Internat. Architektur-Dokumentation, München (2010)
3. Carpo, M. (ed.): The Digital Turn in Architecture 1992–2012. Wiley, Chichester (2013)
4. Sheil, B. (ed.): Protoarchitecture: Analogue and Digital Hybrids. Wiley, Chichester (2008)
5. Sakamoto, T., Ferré, A. (eds.): From Control to Design: Parametric/Algorithmic Architecture. Actar, Barcelona (2008)
6. Knaack, U., Klein, T., Bilow, M., Auer, T.: Façades: Principles of Construction, Second and revised edition. Birkhäuser, Basel, Switzerland, Boston (2014)
7. Böke, J.: ThinkingSkins – towards intelligent acting façades. Paper presented at the Facade 2014 – Conference on Building Envelopes, Lucerne (2014)
8. Klein, T.: Integral Facade Construction: Towards a New Product Architecture for Curtain Walls (2013). doi:10.4233/uuid:f90c1c7e-b6b3-42de-8f17-22cc23981de7
9. Knaack, U., Hildebrand, L., Ashour, A., Pottgiesser, U., Böke, J., Ernst, M.: Prototyping efnMobile. Nai010 Publishers, Rotterdam (2016)
10. Pottgiesser, Hemmerling, Böke (eds.): facade2015 Computational Optimisation. Detmold (2015)

Human, Space and Digitization

Ulrich Nether

1 Introduction

First there is the thesis that digitization has changed our experience and understanding of space.

This article does not start with the planning possibilities that digital tools provide, but instead describes the influence of digitization and computing on space and people in their everyday life.

It introduces a line of thought that imagines a scenario with the goal of gaining insights for dealing with space, with a focus on the tasks of planning. This line of thought is based on the idea that progress and growth in their consumerist expression, according to Western structures of thought, will furthermore be preferred goals in interaction with the world community, and this has consequences for everyone. Alternative scenarios might assume that Eastern thinking is gaining significance and displacing the Modern or fundamentally Religious, or that a profound ecological or economic crisis is occurring. Or, on the way to the world machine, not yet finalized, a butterfly flaps its wings before the machine has had a chance to calculate this—all of this would lead to more or less different developments or formulations.

But the insights from the scenario are transferable, regardless of technology: It might be helpful to consider things more as actors than as objects. Space, then, is less structure than environment with the power to enable things and activities.

U. Nether (✉)
University of Applied Sciences Ostwestfalen-Lippe, Detmold, Germany
e-mail: ulrich.nether@hs-owl.de

M. Hemmerling and L. Cocchiarella (eds.), *Informed Architecture*,
DOI 10.1007/978-3-319-53135-9_11

2 The Essential in the Digital

The Digital is unambiguous; you can count it on your fingers. This means it is existent, concrete, and substantial. The Digital is essence per se.

When we speak of essence, we mean: fundamental to our existence. The essential needs of the human go far beyond pure existence; we need social relationships just as much as we need individual self-esteem and cognitive challenge and we define ourselves through that which we do and for what we strive.

And finally, there are the mythical dreams of the human: According to legend, Prometheus gifted the helpless people the ability to control nature using technology originally the ability of the gods. To make the earth subservient in this way means not only to differentiate ourselves from other organisms nor not only the ability for self-realization, but motivation to transcend. The gods, robbed of their exclusivity, set hurdles and created the seductive Pandora along with her notorious box.

Today, despite adversity, we seem to be on the cusp of discovering the world formula: the analogue and manufacturing are being replaced by digital technology, phenomena transferred into data, saved and processed.

In what we see and experience—and only this is of significance for us—the Digital has become not only an integral part but crucial and even essential; though we ourselves still think we could do without interconnectedness, we hear about the refugees seeking their way to us, that they would rather give away their last shirt than their smartphone.

Digitally formed spaces have long been material, as the word *smombie* (from smartphone and zombie) indicates—the youth un-word of the year 2015 in Germany—or as the worldwide fever surrounding Pokemon Go demonstrates. These new spaces seem generous, they promise immortality, every statement is incorporated into eternal storage, the collective machine-aided human memory. Every time we post a selfie, we transcend ourselves, overcome body and death. In our self-digitization, we ourselves are projects, without presence, oriented towards the future.

Usually, we understand the Digital virtually, as a possible reality. But the stuff dreams are made of is becoming real, dreams are coming true; without any deviation, we produce from data consumer products, furniture, building materials, and in future also food, organs, and organisms are foreseeable to join this list. *AI* is no longer communication between people; the objects themselves are becoming 'intelligent' and networked in the *Internet of Things.* Robots are learning to reproduce themselves, and all of this is systematized in *Big Data*. So, in a few decades, there will likely be nothing more for us humans to do; all functional activities will be done better by machines; but we will have made ourselves immortal by creating something from nature that surpasses us; out of us deficient human beings will arise not just self- and gene-optimized cyborgs in the brave new world, but gods.

This is what is really essential in the Digital, the realization of the human dream: overcoming space and time, paradise on earth, the seventh heaven, nirvana… an infinity in which everything flows and we eventually become superfluous.

3 The Future of Space

The vision is: We no longer have to do inhumane work; all work that can be done by machines will be done by machines. We must no longer suffer; catastrophes can be predicted and the appropriate measures taken; hardships, hunger and sickness nipped in the bud; there is no longer any terrorism or crime. We can access all world knowledge in real time, everything we experience is documented, stored, and processed. Thanks to implanted processors, we can communicate and overcome language and behavioral barriers in the information hegemony of digital rational domination—this is the utopia. The dystopia, on the contrary: we lose ourselves, eliminate ourselves, disappear in one way or another.

Additionally, to think: algorithms do not make mistakes. Algorithms decide; there is no maybe, no hesitation, no failure.

The development of technology as an open system of analysis and synthesis will lead, foreseeably, to the human-machine interface disappearing, to the human in its former understanding of the world disappearing, and to us having to develop a new understanding, new concepts, and new relationships, and to deal with the pursuit of perfection by the world machine we have initiated.

Technological and technical developments cannot be reversed. Therefore, we are dealing with opportunity as a necessity to deal with the insights from the Digital: to get away from the objects, back to the things, away from the object-like conception of space back to a conception that is process- and action-oriented, to use the disappearance of the familiar right up to our conception of the human being and the world to develop something new.

The Digital and its technology is present, hence the question is urgent: who are we and who do we want to be? What should the nature of the new world be and who and what should participate in it and in which way? And what is the nature of the spaces, artefacts, and natural things from whose assembly they are formulated? Which options and relationships will open? Can we use the technical opportunities of the Digital to open systems for an informal open-source world community in which all actors and actants have equal say? And do architects and designers have a part to play in this; those who have expert knowledge of spaces and things? Or do we continue to leave the designing and programming to, in this matter, the less professional euphoric technicians and mission-driven entrepreneurs who are oriented towards it already?

We humans are capable of opposing the nature given to us, to exercise our will over our (out) live artificially; we have will and emotion; we are masters of the art of living and surviving, do-gooders and chronic creatives. Creativity is a faulty circuit and, as a (re)action, a cultivation of the resulting error, innovation the persistent pursuit—a particular quality of human thought and action. Things with self-will, indomitable, capricious, restive, inadequate, failing, foundering, dysfunctional, pointless—let's trust in these, permitting duality, striving not for perfection, but for imperfection, remaining unfinished: To keep the secret means to cultivate the concrete.

For the altered human-space relationship described by the Digital, this would firstly mean to understand space as an environment permanently altered through actions, an environment with individually different important moved and moving actors and actants, and then to derive the in-formation of these spaces and the facts that form them from the relations, relationships, and processes, the things. The insight and challenge of digitization for experts in space, urban planners, architects and interior designers, and for experts in artifacts, designers, is therefore no longer to design structures or objects, but to start from the (inter)action, activity, the processes, the performance, the meanings, and the resonances immanent to these things, to get themselves into the issues behind the things and into a discussion of human and things—to compose possibilities for improvisation and performance.

References

1. Flusser, V.: Dinge und Undinge. München (1993)
2. Latour, B.: Das Parlament der Dinge. Naturpolitik, Frankfurt am Main (2001)

Part II
Case Studies

Design-Build Projects in Academia

Marco Hemmerling

1 Applied Research as Teaching Method

Digital technologies altered the field of architecture and the architectural profession significantly from design to production. Even though CAD as drawing tool became a standard throughout the last decades, both in architecture schools and practices, there is still a huge potential of computational technology waiting to be activated for architectural production processes. However, to get in-depth understanding of the capability of digital applications, it requires not only time and the ability to operate these tools but foremost a broad expertise in architectural design methodology comprising of Computer-Aided Design (CAD), parametric design strategies, Building Information Modelling (BIM), simulation tools, and digital fabrication technologies plays a significant role in the evolution of contemporary architecture. This situation calls for a redefinition of the architect's education. It requires (future) architects to become familiar with computational processes and to understand the underlying principles. Recognizing the potential of this development, new technologies and applications promise more than just new design possibilities. Above all, it puts the architect back to their central role within the architectural production process, like the building master, being responsible from the initial idea to the completed building. This prospect becomes apparent in many areas of the architectural process through the ability to respond directly to design iterations, by saving time through automation, by increasing economic viability, by reducing the potential sources of error, and through the ability to directly transfer the digital design into physical results by the means of latest CAD/CAM (Computer Aided Manufacturing). Even though CAD-modules are implemented in most of the Bachelor and Master courses in architecture schools, it is just one in between many

M. Hemmerling (✉)
Computational Design in Architecture, Faculty of Architecture,
Cologne University of Applied Sciences, Cologne, Germany
e-mail: marco.hemmerling@th-koeln.de

M. Hemmerling and L. Cocchiarella (eds.), *Informed Architecture*,
DOI 10.1007/978-3-319-53135-9_12

other subjects of the whole curriculum. Taking into account the fast development and variety of digital tools and the complexity of their applications on one hand and the demands for a reflective methodology that allows for an intelligent and appropriate design strategy on the other, the need for an in-depth understanding and a mutual integration of this field in the architectural discourse is absolutely mandatory.

Against this background, the second part of the present volume documents a range of academic projects carried out at the Detmold School of Architecture and Interior Architecture, East-Westphalia University of Applied Sciences as well as at the Politecnico di Milano from 2008 to 2017. This chapter collects a compilation of visionary, playful, and experimental ideas that were not only realized by the use of digital tool and computational strategies, but above all, with an approach of encouragement, inspiration, and imagination.

The selected case studies reflect the previously discussed theoretical background and exemplify the various aspects in specific projects—from conceptual design to prototypical building. The projects range from one week workshops to entire academic year modules, carried out on Bachelor-, Master- and Postgraduate-level. The kaleidoscopic documentation showcases not only the results but foremost, tries to delineate the underlying teaching methods and findings throughout the process. As such the case studies should serve as an inspirational source for students, lecturers, and architects and encourage the development and transformation of these concepts. Moreover, the documentation is intended to foster the discussion about the potentials, conditions, and constraints of computational design and construction strategies in architecture.

Twister

Marco Hemmerling, Ulrich Knaack, Jens-Uwe Schulz, David Lemberski, Simon Aksal, Yalcin Dündar, Benjamin Gerner, Jakob Heining, Matthias Kemper, Rüdiger Kiene, Jan Klassen, Manuel Welsky and Thomas Wirtel

Detmold School of Architecture and Interior Architecture
East-Westphalia University of Applied Sciences, Germany
(2008)

The academic project "Twister" was set-up with the aim to generate a spatial design, which is inspired by fascinating structures from nature and at the same time based on traceable geometric principles and rules. Moreover, the challenge was not only to generate a complex design, but also to build it in scale 1:1.

In the beginning, the inspiration of a natural structure was translated into a mathematical model. Based on that 3D-Model, the generative elements of the geometry—like points, edges and surfaces—where transformed in order to achieve self-similar models with slightly different properties. In a process of selection, the models where categorized in relation to their spatial and structural qualities and their ability to be represented in an appropriate geometric model for the production of large scale elements.

M. Hemmerling
Cologne University of Applied Sciences, Cologne, Germany

M. Hemmerling (✉) · U. Knaack · J.-U. Schulz · D. Lemberski · S. Aksal · Y. Dündar · B. Gerner · J. Heining · M. Kemper · R. Kiene · J. Klassen · M. Welsky · T. Wirtel
Detmold School of Architecture and Interior Architecture,
East-Westphalia University of Applied Sciences, Detmold, Germany
e-mail: marco.hemmerling@th-koeln.de

U. Knaack
e-mail: u.knaack@tudelft.nl

J.-U. Schulz
e-mail: jens-uwe.schulz@hs-owl.de

D. Lemberski
e-mail: david.lemberski@hs-owl.de

U. Knaack
Techincal University Delft, Delft, The Netherlands

M. Hemmerling and L. Cocchiarella (eds.), *Informed Architecture*,
DOI 10.1007/978-3-319-53135-9_13

The design process can be described as a sequence of inspiration, abstraction, variation, and selection. The structure of a winding nautilus-snail served as an initial inspiration for the design process. By analyzing and rationalizing the geometry of this natural structure, a digital 3D-model was developed that interconnects different arcs to an intertwining surface. The radiuses of the arcs increase from the center to the outside of the structure while the arcs are rotated by 10° in XY-direction around the center. The spatial performance can be described as an in-between of directional and centered space. The shape seemed to be ideal to function as an exhibition space that allows a seamless transformation of guided circulation while entering and leaving the space to focused concentration while being in the center of the structure.

The next steps in the process comprised the investigation of different construction concepts for the exhibition pavilion. The idea of a textile skin—supported by glass-fiber-rods, which define the geometry of the arcs—was finally chosen as a structural concept. The tent-like structure didn't only respond perfectly to the geometric principles of the digital model, but also offered a simple production and construction method for the pavilion. Based on the translucency of the textile membrane, the illumination of the structure by night became a new aspect of the design. For the building process of the low-budget project, a common tent structure based on carbon-fiber rods for the arcs and a textile membrane for the minimal surfaces spanning between the rods was chosen.

Within the further planning process, the conditions of production (fixtures, element sizes etc.) and the material properties were investigated and implemented into the digital model. By using the principles of reverse engineering, the digital 3D-geometry was adapted to the parameters of production. The geometric deviations of the rods in relation to the digital model of the arcs was analyzed in a test-assembly of each arc. With the help of Photogrammetry, that relates the lines of the digital model to the orthoscopic photo of the rods, the physical geometry of the rods was ascribed to the Rhinoceros-Model. Based on this analysis, the digital model was adapted to the material conditions from the survey of the rods. The blanks and sizes for the rods and the membrane where finally taken from the updated model in Rhinoceros into the realization process.

The double curved geometry of the minimal surfaces was unfolded by using approximation methods after generating polysurfaces from the NURBS-model. The different shapes of the developed surfaces where then used as a layer for the tailoring of the membrane. The dialogue between digital and physical methods resulted in a precise description of the building parts. The assembly of the exhibition pavilion "Twister" was finally based on rather traditional methods of membrane constructions using sewing-technologies and knot-fixtures. After setting out the rods, that defined the basic geometry, the tailored membrane was applied to the line-grid of the rods from the bottom to the top (Figs. 1, 2, 3, 4, 5, 6, 7 and 8).

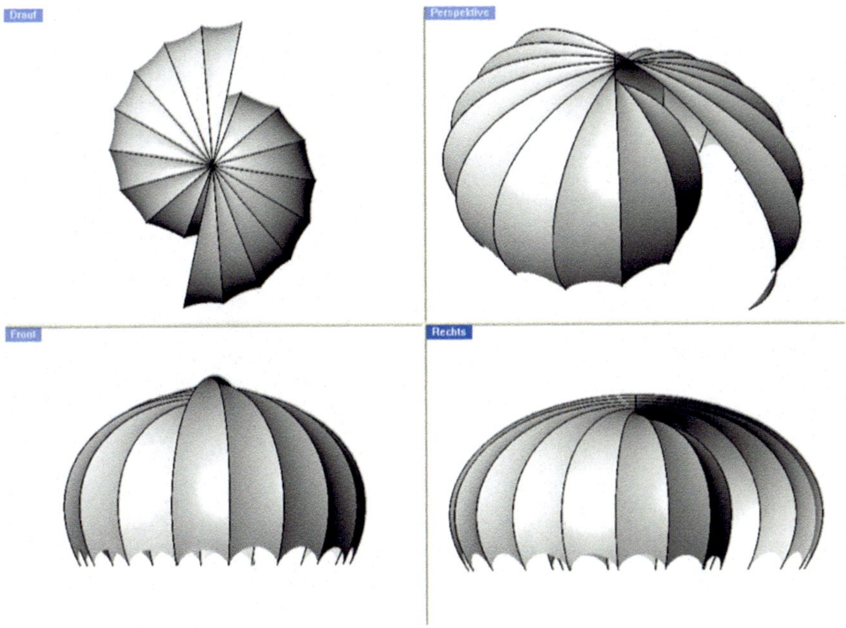

Fig. 1 Four panel projection of the digital design model

Fig. 2 Twister rendering

Fig. 3 3D-printed scale model

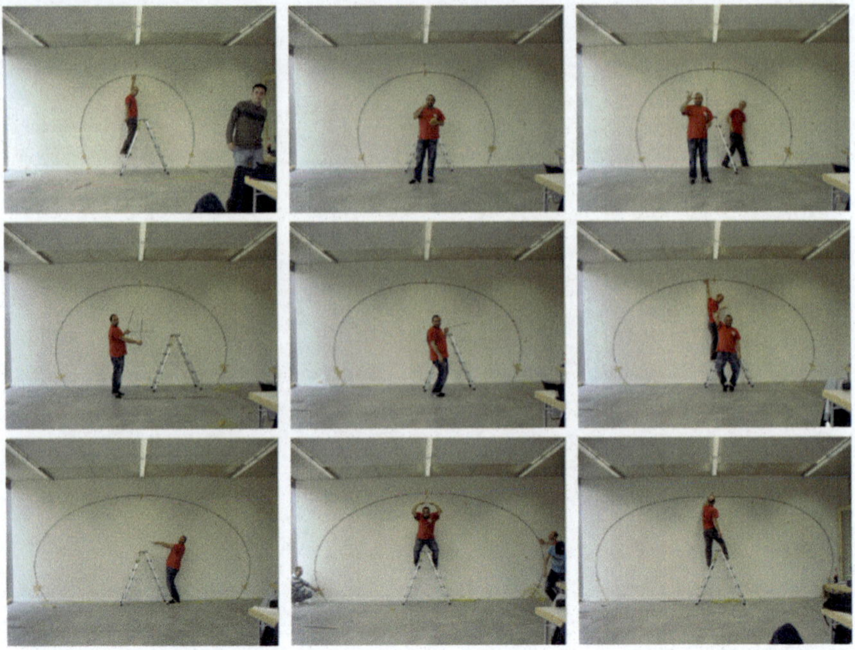

Fig. 4 Mock-up for the photogrammetry of the various arcs

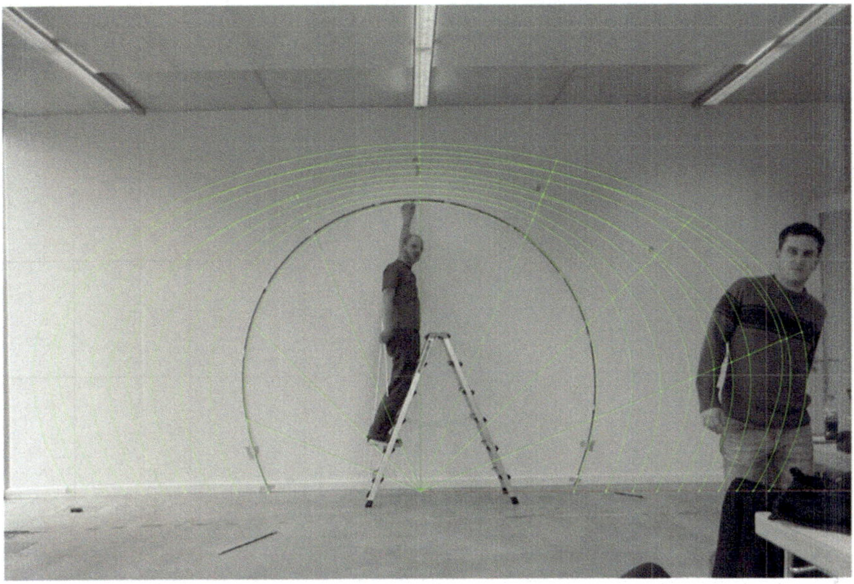

Fig. 5 Superimposition of the digital geometry and the physical mock-up

Fig. 6 Assembly of the arcs on the University campus Emilie in Detmold

Fig. 7 Central connection detail

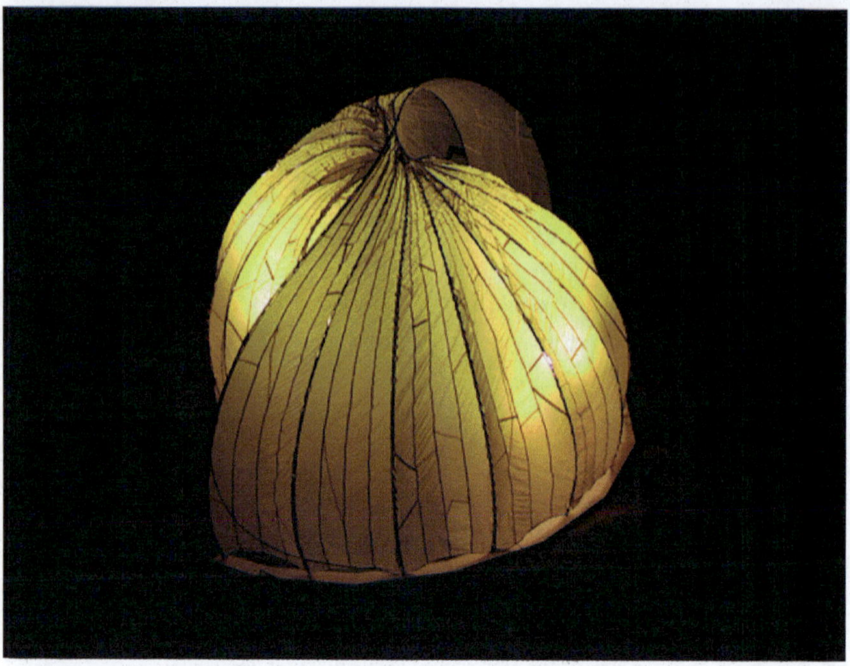

Fig. 8 Twister pavilion by night

SunSys

Marco Hemmerling, Ulrich Nether and Jens Böke

Detmold School of Architecture and Interior Architecture
East-Westphalia University of Applied Sciences, Germany
(2008, 2014)

SunSys focused on an all-embracing design approach, implementing and connecting all necessary information in a computational design process. Therefore, the early integration of optimization parameters, regarding structural performance, physical properties and material specification as well as aspects of fabrication to inform the architectural design and construction were key aspects of the project.

In order to match the overall requirements, various input parameters were developed and integrated into a parametric definition. They were established by analyzing onsite conditions using GPS tracking systems, image acquisition series, and solar radiation analysis. In particular, existing patterns and structures of movement onsite, usage timeframes, and alignment in relation to the course of the sun and to changing patterns of shade were used as source data for the design strategy. The form-finding process developed generatively while the values for the structures, graduated from the succession of open and enclosed sections, were derived from the findings of the time-usage analysis and the sun analysis. The profiles of the different usage areas, which form the pavilion's geometry, were created with reference to different seat positions related to furniture—like a desk chair, bench, lounge- or deckchair.

After a selection process from multiple variations, the final version of the pavilion was to be constructed from planar elements that are connected at an angle

M. Hemmerling (✉) · U. Nether · J. Böke
Detmold School of Architecture and Interior Architecture,
University of Applied Sciences Ostwestfalen-Lippe, Detmold, Germany
e-mail: marco.hemmerling@th-koeln.de

U. Nether
e-mail: ulrich.nether@hs-owl.de

J. Böke
e-mail: jens.boeke@th-koeln.de

M. Hemmerling and L. Cocchiarella (eds.), *Informed Architecture*,
DOI 10.1007/978-3-319-53135-9_14

of 90° in order to simplify the assembly process. While all elements differ in shape, the CNC-fabrication of the building parts is based on a unified principle. The individual building parts, which also constitute the connective elements, are based on a parametric definition that incorporates all parameters but also permits endless variations, in terms of combination, allowing for an ideal form with a high degree of diversity and precision. Previously constructed surface objects define the overall shape of the pavilion while the planar elements are programmed by a parametric definition in Rhinoceros/Grasshopper. After this system is set up, variations can be applied easily. For instance, by scaling the dimensions of the guiding elements following collected parameters, such as, the distance towards an attractor-point. In actuality, the form-definition of the SunSys-pavilion project depends on environmental parameters (solar radiation) following this logic.

SunSys should be understood as a prototype to achieve diversity and flexibility by computational efficiency. Given a different set of parameters, the design strategy would produce different overall shapes, functions, and usage profiles while the underlying geometric principle and fabrication method would remain the same. It is an illustrative example of how to benefit from parametric design and the resulting simplification of production and assembly processes without the need to post-rationalize geometric complexity (Figs. 1, 2, 3, 4, 5 and 6).

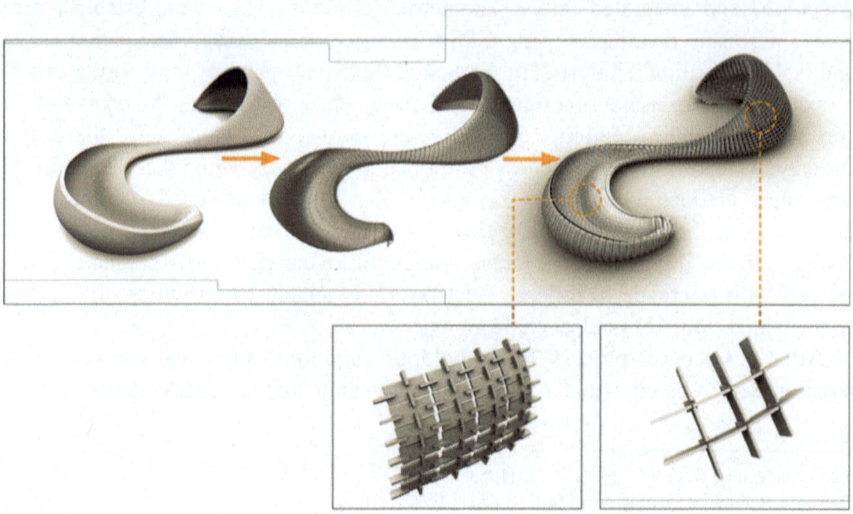

Fig. 1 Design development: From the overall shape to the connection detail

Fig. 2 Sun analysis and shadow study

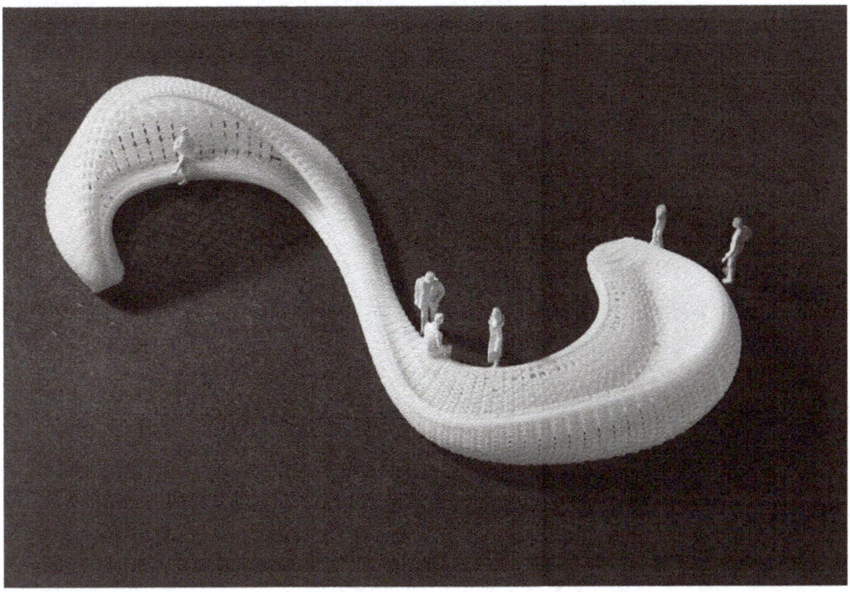

Fig. 3 3D printed scale model

Fig. 4 SunSys rendering

Fig. 5 SunSys rendering

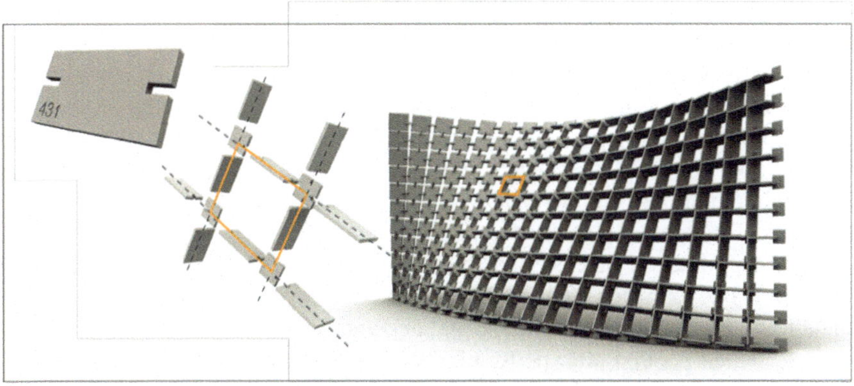

Fig. 6 Connection detail and variation of the systems

In 2014, the system was developed further towards a design-build project in scale 1:1 as a proof of concept. The aim of the recent reconsideration of the draft was to transfer the digital model into a buildable construction: A five meters long and two meters high wall, which was implemented as an experimental construction on the campus Emilie in Detmold. It was developed by the application of digital design and manufacturing methods. Planar elements are assembled in a plug-in system and generate a freeform geometry. Variations in the dimension of the individual elements enable different degrees of wall-openings.

The revised Grasshopper definition evaluates the initial surfaces and subdivides them by their UV-coordinates. In the following step, each coordinate is equipped with a plane, oriented to the normal-Vector of the surface. As an in-between result, a rectangular grid of perpendicular planes describes the pavilion's geometry. To achieve the system logic, three different groups of planar elements have been generated:

- The ones, which connect both other groups by being perpendicular on the guiding surfaces.
- A group of elements, connecting the perpendicular elements in the U-direction, by rotation around their x-axis.
- A group of elements doing the same for the V-direction by rotation around their y-axis.

The rotation of the elements is defined by orienting every second plane, counting in both directions. The elements are generated as rectangular curves onto the planes. The perpendicular flat elements are created as guiding-elements; their dimension is for each direction half of the distance to the next evaluated point. It is important for the precision of the result to follow these guidelines, otherwise, the planar interpretation of the double-curved geometry leads to uncontrollable tolerances by just subdividing the distances. In order to generate the rectangle curves of both other element-groups, a mid-point is generated on the rectangle-lines for each direction, being the start or endpoint of a line, connecting to the midpoint of the next rectangle-line in that direction. By extruding the lines in the direction of their plane x- or y- direction, the rotated elements can be generated as rectangular surfaces. Accordingly, the distance of the extrusion is defining the geometry depth of the system. For the connection of the elements, each element has to be cut by half the distance of its intersection with the neighboring element (waffle-principle). Assembling both half cut elements will finally perform the full intersection of the elements. Through a Boolean operation, the rectangles are extruded to solids.

As the production process and the chosen material require the calculation of tolerances, two extrusions have to be performed: one without adding the tolerance, to define the final geometry element, and one with an added tolerance to be just the cutting element, erased after performing the cut to its neighbor. In order to start a CNC-production process from planar panels, the geometry of the resulting 3D-Model has to be unrolled. In addition, it was necessary to give identities to the 3D-objects that can also be milled into the physical pieces to keep control of the production and assembly process (Figs. 7, 8, 9, 10, 11, 12, 13, 14 and 15).

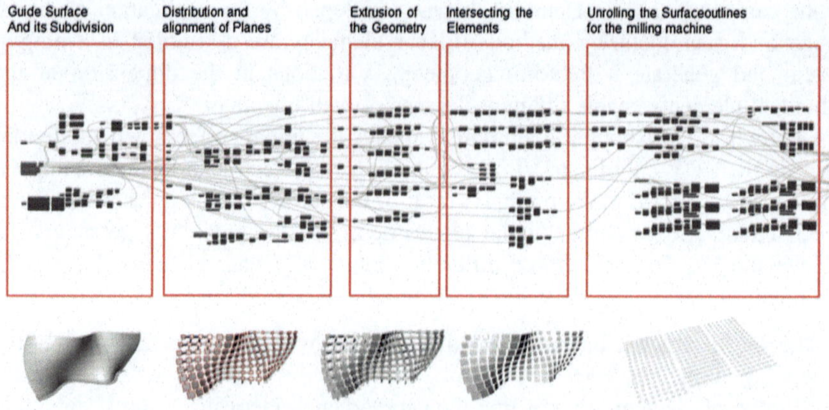

Fig. 7 Grasshopper definition of the wall system

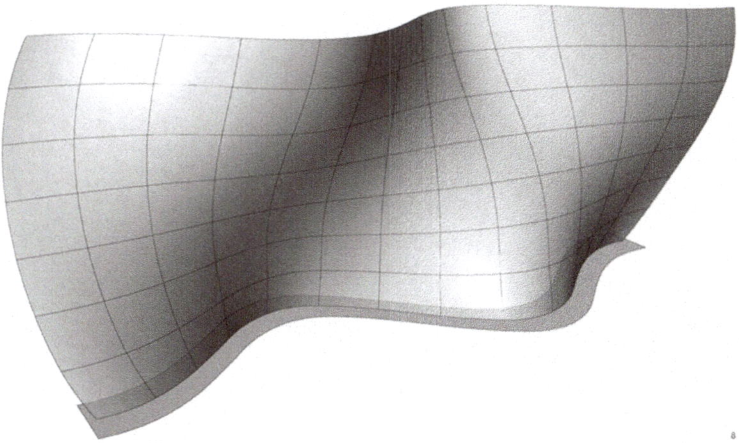

Fig. 8 Double-curved guiding surface

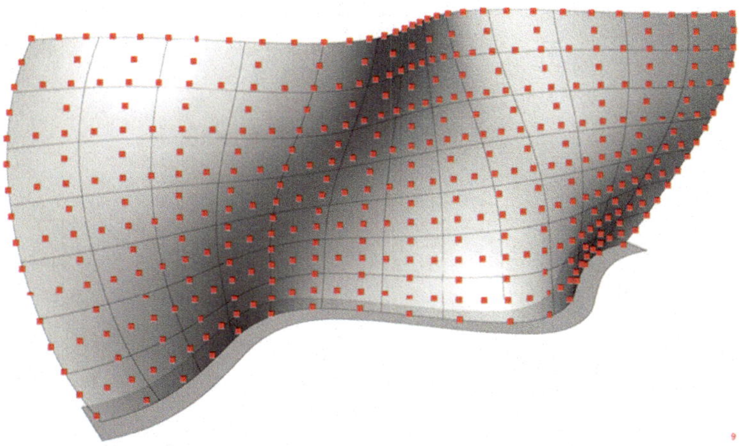

Fig. 9 Subdivision of the surface

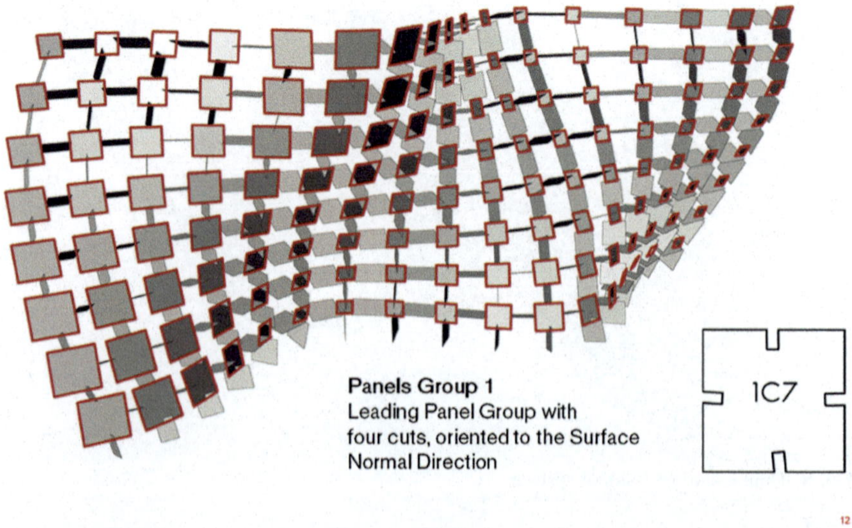

Fig. 10 Distribution and alignment of the panels–group 1

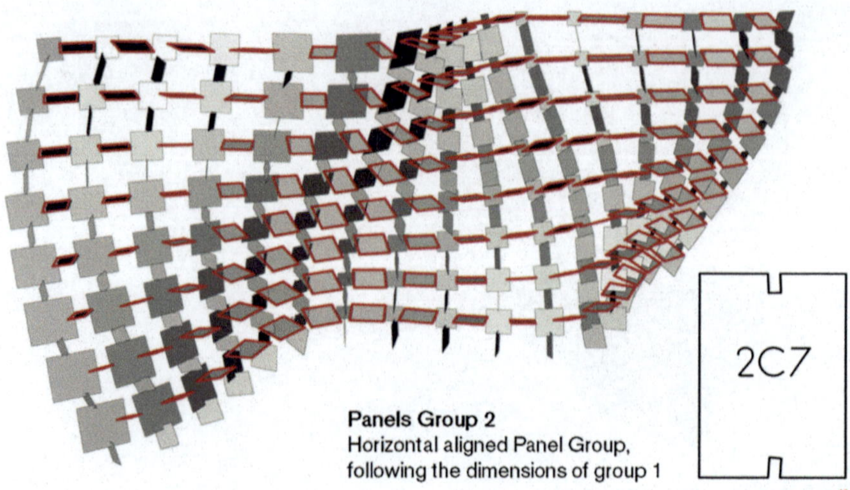

Fig. 11 Distribution and alignment of the panels– group 2

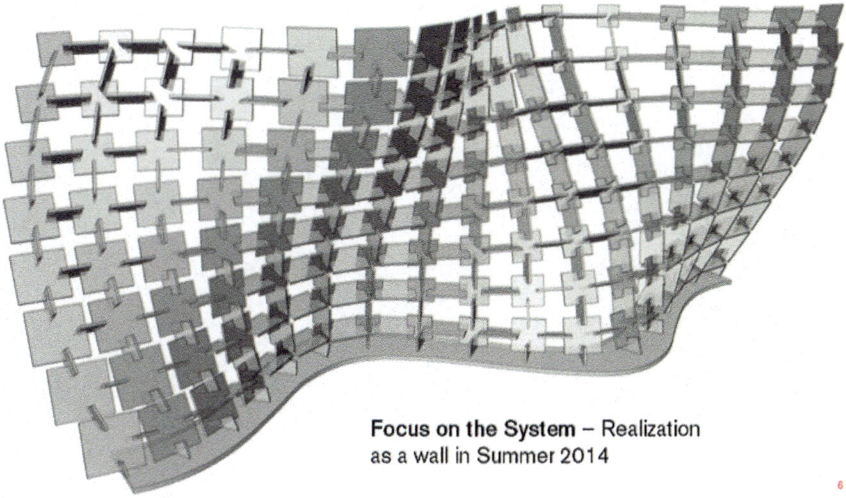

Fig. 12 Extrusion of the panels according to the material thickness

Fig. 13 Realized wall system on the Campus Emilie in Detmold

Fig. 14 Realized wall system on the Campus Emilie in Detmold

Fig. 15 SunSys connection detail

Boxel

Marco Hemmerling, Matthias Michel, David Lemberski, Guido Brand, Henri Schweynoch, Guido Spriewald, Elena Daweke, Lisa Hagemann, Bernd Benkel, Thomas Serwas, Michael Brezina, Samin Magriso, Caroline Chi Zhang and Michelle Layahou

Detmold School of Architecture and Interior Architecture East-Westphalia University of Applied Sciences, Germany (2010).

The design for the pavilion *Boxel* was carried out by a competition among Bachelor students. The building is based on a minimal surface and consists of 1950 beer crates that are organized along a free-form geometry. The idea to take a product from a different context and to use it as a non-standard building material to generate a new perception of the module convinced the jury. Henri Schweynoch's design was also chosen due to its expressive spatial appearance and fascinating relation between the rigid module and the curved shape of the pavilion. The usual vertical stacking of the boxes is dissolved to a free organization that could basically take any shape. The geometry evolves from three support points and forms three arcs that opens up towards the main buildings of the Campus Emilie and react at the same time to the relevant lines of sight.

Going in the process further, the 3D-Model was programmed in *Rhinoceros/ Grasshopper* to define the overall shape in relation to the distribution and position of the orthogonal module along the surface. By using this flexible design tool, variations of the design were carried out, taking into account the spatial appearance,

M. Hemmerling (✉) · M. Michel · D. Lemberski · G. Brand · H. Schweynoch · G. Spriewald · E. Daweke · L. Hagemann · B. Benkel · T. Serwas · M. Brezina · S. Magriso · C.C. Zhang · M. Layahou
Detmold School of Architecture and Interior Architecture, East-Westphalia University of Applied Sciences, Detmold, Germany
e-mail: marco.hemmerling@th-koeln.de

D. Lemberski
e-mail: david.lemberski@hs-owl.de

G. Brand
e-mail: guido.brand@hs-owl.de

M. Michel
Imagine Structure, Frankfurt, Germany
e-mail: michel@imagine-structure.com

M. Hemmerling and L. Cocchiarella (eds.), *Informed Architecture*,
DOI 10.1007/978-3-319-53135-9_15

the connection of the adjacent crates, and the optimization of the structural behavior. For a deeper understanding of the statics of the model, the students teamed up with the department of building engineering to simulate and analyze the internal forces, deformations, and support reactions of the shell using the structural engineering software *R-Stab* and *R-FEM*.

In order to define the construction concept and the detailing of the connection, several static load tests were made to understand the structural behavior of the unusual building material, especially since the empty beer crates were not stacked onto but freely organized next to each other.

Parallel to the digital planning process, a series of load tests were realized in the university's laboratory of material research at the departments of building engineering and material science. Three different breadboard constructions were set-up for the material and construction tests in order to understand the structural behavior of the module and the connection (bending, shearing, diagonal load).

Boxel was erected in only one week by the students and served as a scenic background for the end of semester party and during the international summer at the University in Detmold. The beer crates which, after being ten years in use, were supplied by the local brewery and will be recycled when the pavilion is being disassembled (Figs. 1, 2, 3, 4, 5, 6, 7, 8, and 9).

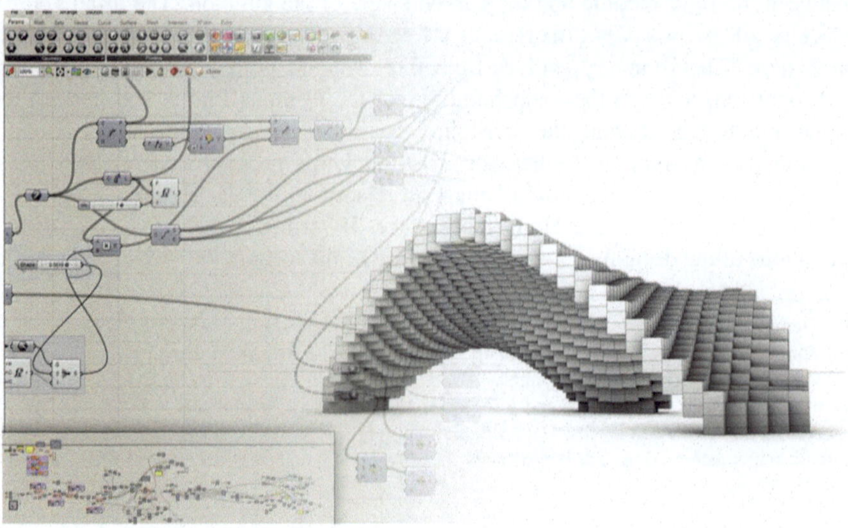

Fig. 1 Grasshopper definition of the geometry

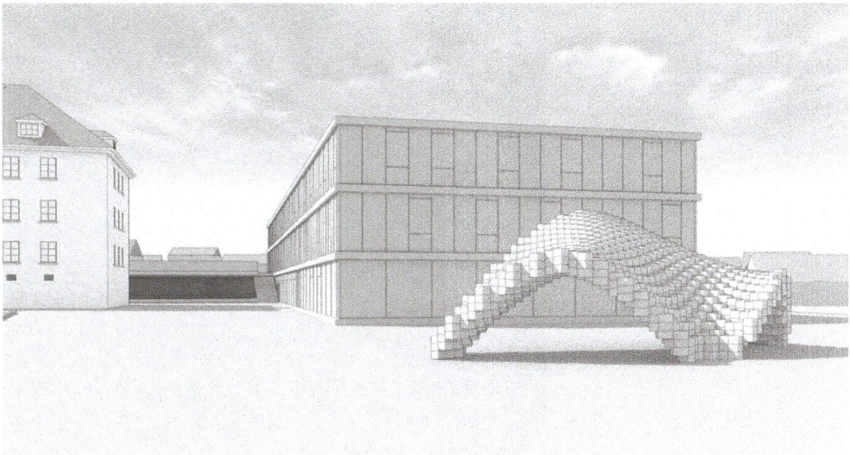

Fig. 2 Visualization of the structure on the Campus Emilie in Detmold

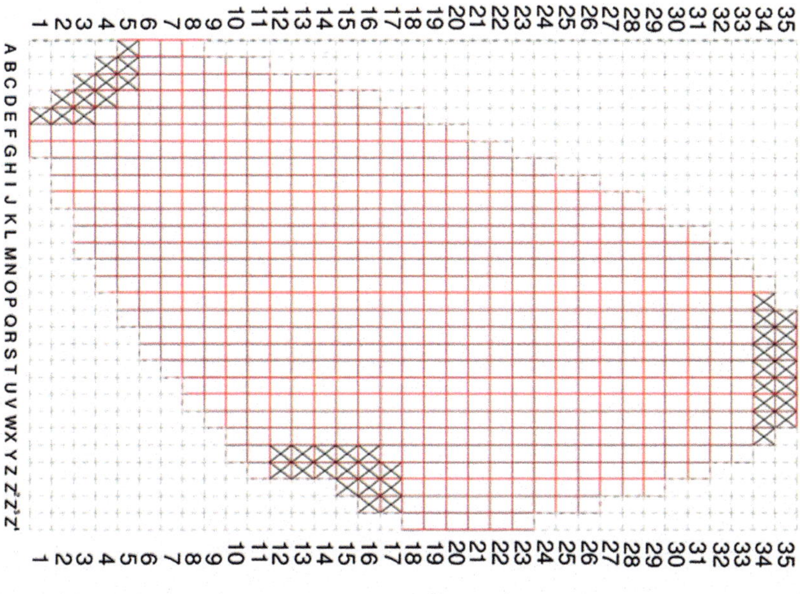

Fig. 3 Grid system of the Boxel pavilion with marked supports

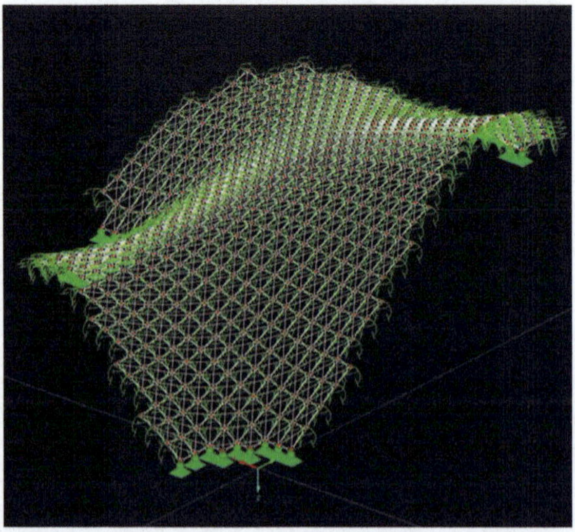

Fig. 4 Structural analysis

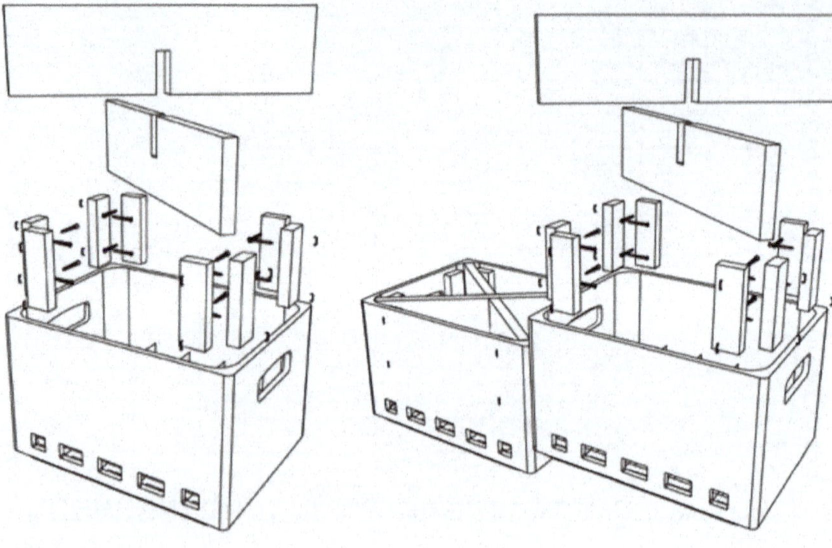

Fig. 5 Construction principle and bracing for the boxes

Fig. 6 Assembly of the structure on site

Fig. 7 Inauguration of the final structure on the Campus Emilie in Detmold

Fig. 8 View of the structure with a 15 m span and a total height of 5 m

Fig. 9 Boxel pavilion by night

Design to Production

Marco Hemmerling, Swantje Kühn and Frank Püchner

Detmold School of Architecture and Interior Architecture
East-Westphalia University of Applied Sciences, Germany
(2010).

This architectural diploma dissertation by Frank Püchner took, as its starting point, the possibilities for producing curved surfaces using computer-assisted tools. He investigated a number of construction systems in relation to digital construction variants while also implementing them in physical models. The first step was to produce the base geometry, a translation surface composed of sine curves, on the computer. This then served as the starting point for the investigation of a variety of production and construction variants.

During the next phase—the actual production phase—an issue emerged which warranted that changes and additions to the digital data model (in terms of values, materials, positions, details and other factors) would be required in order to implement the digital model using computer-controlled machines. In particular, it was necessary to integrate readings on production tolerances, static requirements and connection and join parameters, and the properties of the materials used (various wood types, hard foam) into the dataset involved in the transition from planning to production.

To avoid having to produce the 3D model again from scratch as a result of these optimisation measures, a parametric model permitting adaptation was developed using Rhinoceros 3D software. The digital model was duly parameterised with the aid of the Grasshopper plug-in, enabling modifications to be applied to the model in a holistic way. This allowed all the building blocks in the process to be simulta-

M. Hemmerling (✉)
Cologne University of Applied Sciences, Cologne, Germany
e-mail: marco.hemmerling@th-koeln.de

M. Hemmerling · S. Kühn · F. Püchner
Detmold School of Architecture and Interior Architecture,
East-Westphalia University of Applied Sciences, Detmold, Germany
e-mail: swantje.kuhn@hs-owl.de

M. Hemmerling and L. Cocchiarella (eds.), *Informed Architecture*,
DOI 10.1007/978-3-319-53135-9_16

neously controlled and altered by means of a single central file. Grasshopper's visual programming also provided a template for practical implementation of the series of experiments. This resulted in a digital process chain capable of serving as the basis for a continuous development process, from the creation of the concept geometry to the refining of the 3D model to the realisation. The parameterising process allows any construction approach to be directly adapted and optimised as well as to be evaluated in design terms.

In addition to the horizontal layering, triangulation, and insertion raster system (waffle-grid) production principles, Frank Püchner's architectural diploma dissertation incorporated the ZIP principle, as developed by Christoph Schindler,[1] programmed as a functional dataset. As a result, these digital tools do not relate solely to a single geometry—they also allow for the production of diverse single- and double-curved geometries (Figs. 1, 2, 3, 4, 5, and 6).

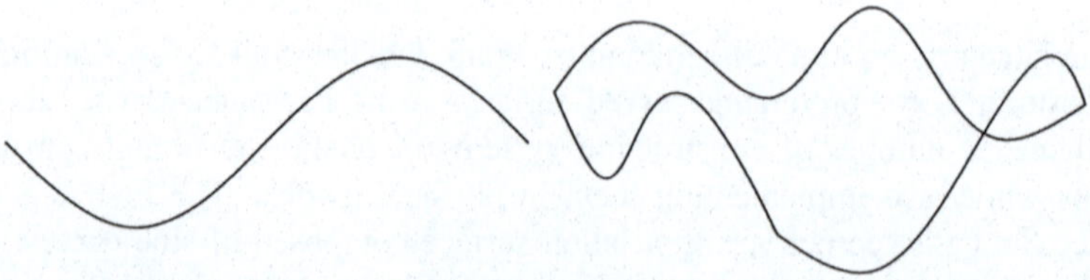

Fig. 1 Generating outlines of the double-curved surface

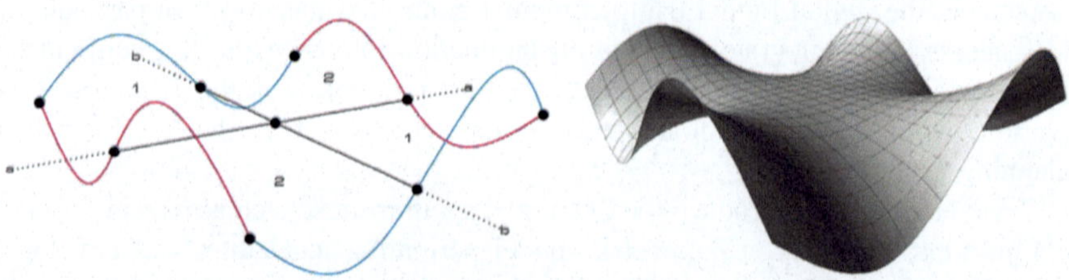

Fig. 2 Geometric definition of the translational surface

[1]Schindler C., ZIP-Shape, Proceedings of the 26th eCAADe-Conference, Antwerp, 2008.

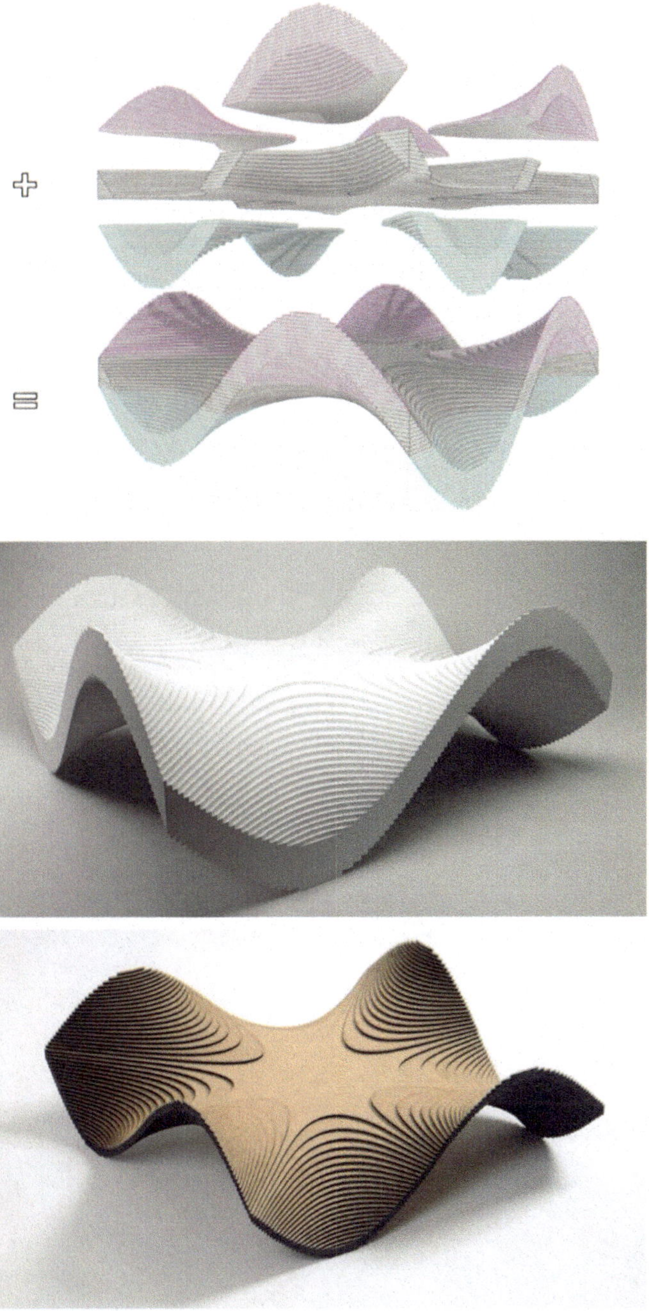

Fig. 3 Horizontal layering: Principle, digital and physical model (top–down)

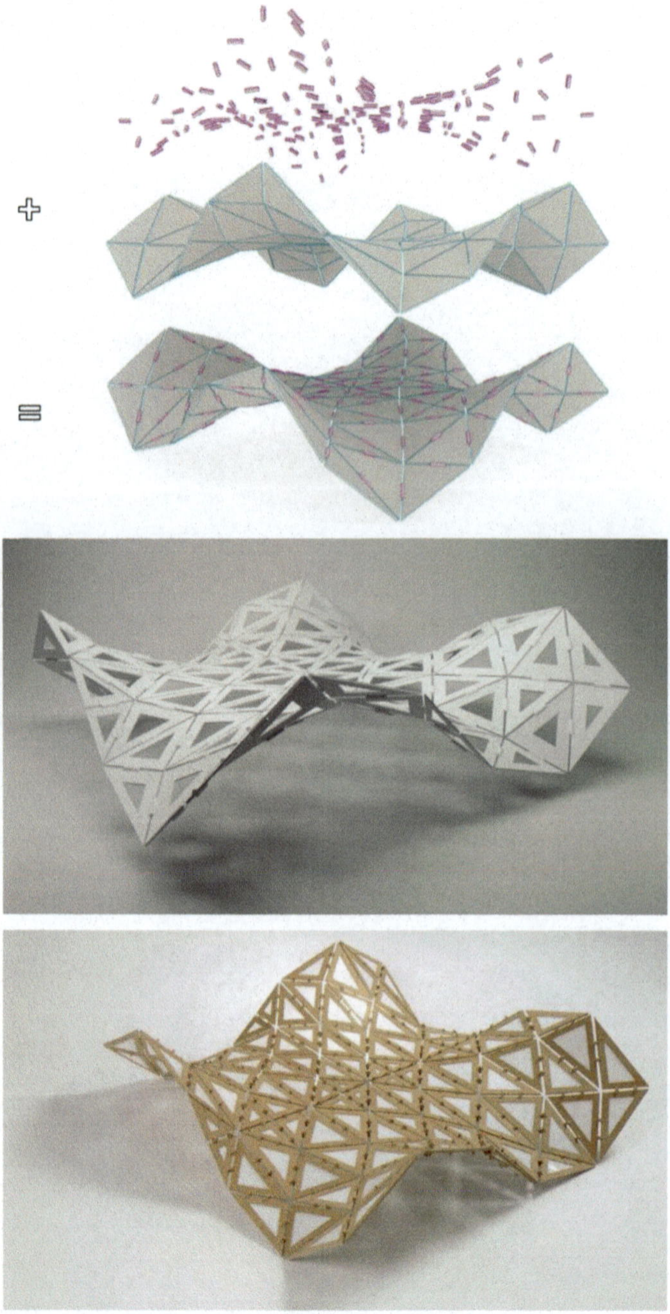

Fig. 4 Triangulation: Principle, digital and physical model (top–down)

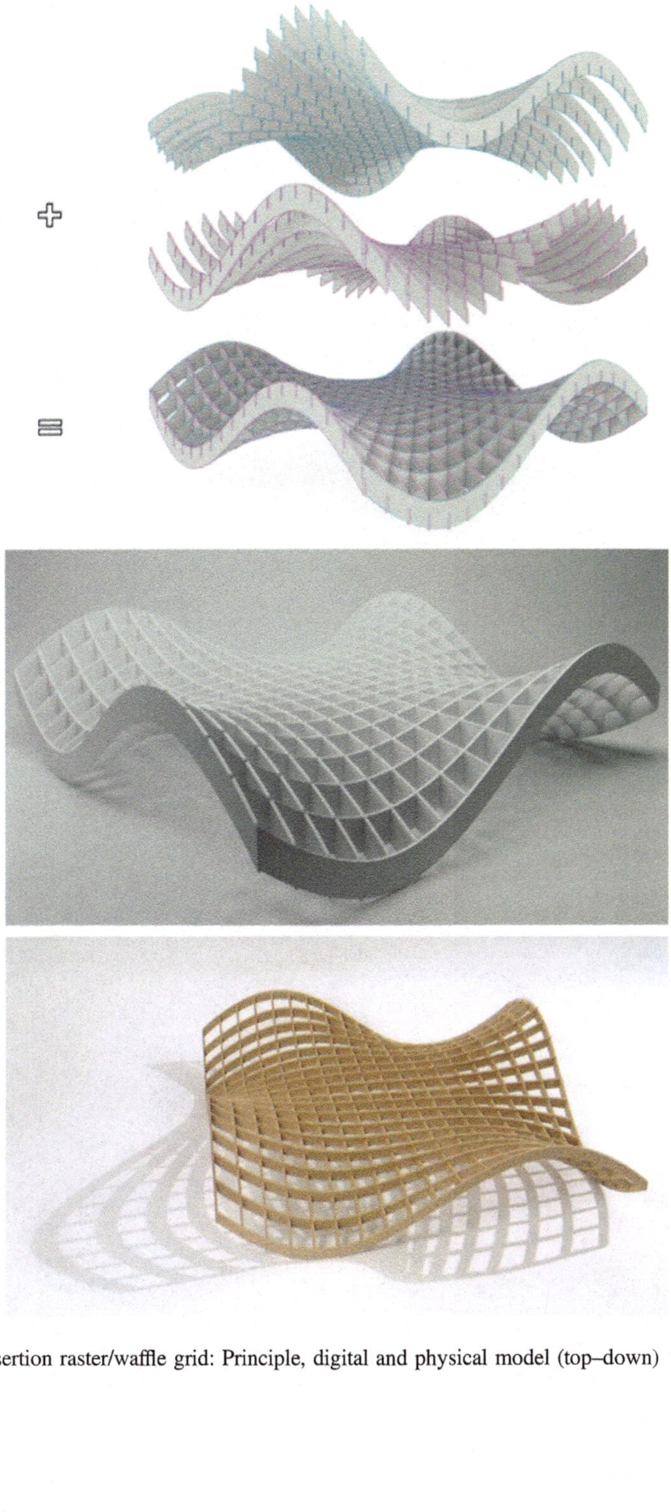

Fig. 5 Insertion raster/waffle grid: Principle, digital and physical model (top–down)

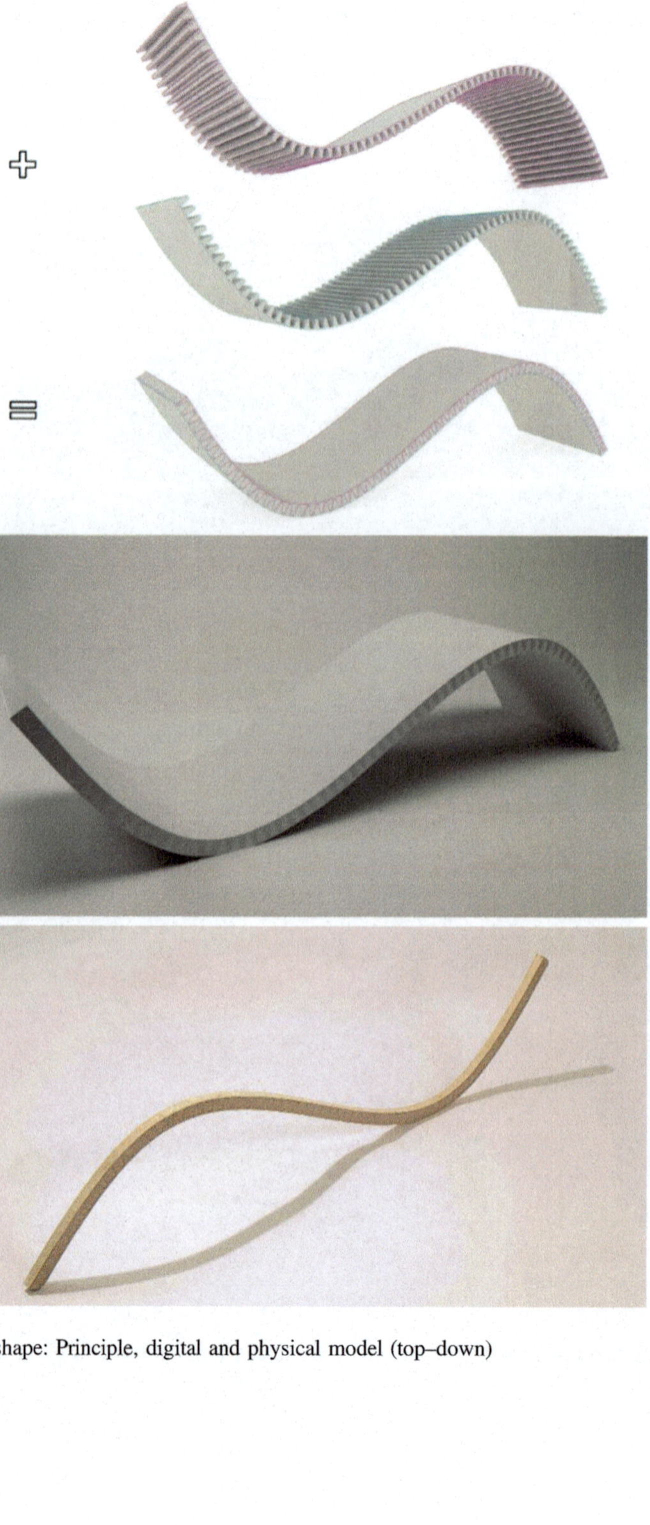

Fig. 6 ZIP-shape: Principle, digital and physical model (top–down)

Sparkler

Marco Hemmerling, Holger Hoffmann, Matthias Michel, David Lemberski, Guido Brand, Miriam Johnson, Fatima Ibrahimi, Janina Pörtner, Pia Hartmann, Felicitas Wendler, Jannik Flöttmann, Maria Moormann, Jan Baumgartner, Manuel Münsterteicher, Matthias Joachim, Wilrun Griemert, Ann-Kathrin Kahmann, Inga Sonntag, Viktoria Padberg, Julia Scheppke, Tülin Zümre, Janine Wilkop, Inna Metche, Esther Jablotschkin, Johanna Stüve, Stefanie Bröckling, Martina Driller, Irina Kraus, Olga Schukow, Lea Mattenklodt, Christoph Strugholtz, Juliane Meyer and Judith Woker

Detmold School of Architecture and Interior Architecture East-Westphalia University of Applied Sciences, Germany (2011).

Sparkler is a three-dimensional interpretation of the well-known study of proportions by Leonardo da Vinci which shows the Vitruvian man within a circle and a square. Seen from the outside, this experimental pavilion resembles an angular crystal while its interior conforms to a perfect spherical geometry. These two basic

M. Hemmerling (✉) · H. Hoffmann · M. Michel · D. Lemberski · G. Brand · M. Johnson · F. Ibrahimi · J. Pörtner · P. Hartmann · F. Wendler · J. Flöttmann · M. Moormann · J. Baumgartner · M. Münsterteicher · M. Joachim · W. Griemert · A.-K. Kahmann · I. Sonntag · V. Padberg · J. Scheppke · T. Zümre · J. Wilkop · I. Metche · E. Jablotschkin · J. Stüve · S. Bröckling · M. Driller · I. Kraus · O. Schukow · L. Mattenklodt · C. Strugholtz · J. Meyer · J. Woker
Detmold School of Architecture and Interior Architecture,
East-Westphalia University of Applied Sciences, Detmold, Germany
e-mail: marco.hemmerling@th-koeln.de

H. Hoffmann
e-mail: hoffmann@uni-wuppertal.de

M. Michel
e-mail: michel@imagine-structure.com

D. Lemberski
e-mail: david.lemberski@hs-owl.de

G. Brand
e-mail: guido.brand@hs-owl.de

M. Hemmerling
Cologne University of Applied Sciences, Cologne, Germany

M. Hemmerling and L. Cocchiarella (eds.), *Informed Architecture*,
DOI 10.1007/978-3-319-53135-9_17

forms are interconnected by the extended edges of a regular Archimedean form—a blunted icosahedron—in such a way that the resulting spatial sculpture appears as an internally harmonious shape.

In the first stage of the designing process, a parametric geometrical model was constructed with the aid of the Rhinoceros/Grasshopper 3D modelling software. This geometrical model allowed the base geometry for both the interior and the exterior—a sphere and a block as space-enclosing volumes—and the various interface surfaces that ultimately create the form of the body to be placed in relation to each other. This 3D model, created to serve as a designing machine, made it possible for a large number of variations on the basic idea to be produced quickly, evaluated, and compared. Varying the sections on the external sides allows the pavilion to occupy seven different positions. As well as expanding the possibilities for use, this also alters the appearance of the sculptural structure in ways that are dependent on angles of perspective and on the contact area selected.

Additional information relating to production was added to the parametric model in order to prepare for production of the individual components using CNC technologies. In addition to optimised cutting to the size of the panels (known as nesting), integrating the material thicknesses and the required tolerances for assemblage, logistical aspects were a factor in the construction, deconstruction, and reconstruction of the digital model. In addition to helping to visualise the design spatially, a scale wood model created using the laser cutter (1:10) assisted in the preparation for and testing of the subsequent assemblage process.

Because the prototypical structure can be developed in an idealised process, allowing a fundamental understanding of the methods presented, the pavilion's typology presents a suitable field of experimentation for new spatial concepts. The 90 different geometrical multiplex panel formats from which Sparkler is composed are connected by over 120 laser-cut force-fitted steel connections. This means that the structure, which is based on a "buckyball", shows statically optimised load-bearing behaviour, also permitting serial production of the identically shaped steel connectors (Figs. 1, 2, 3, 4, 5, 6, 7, and 8).

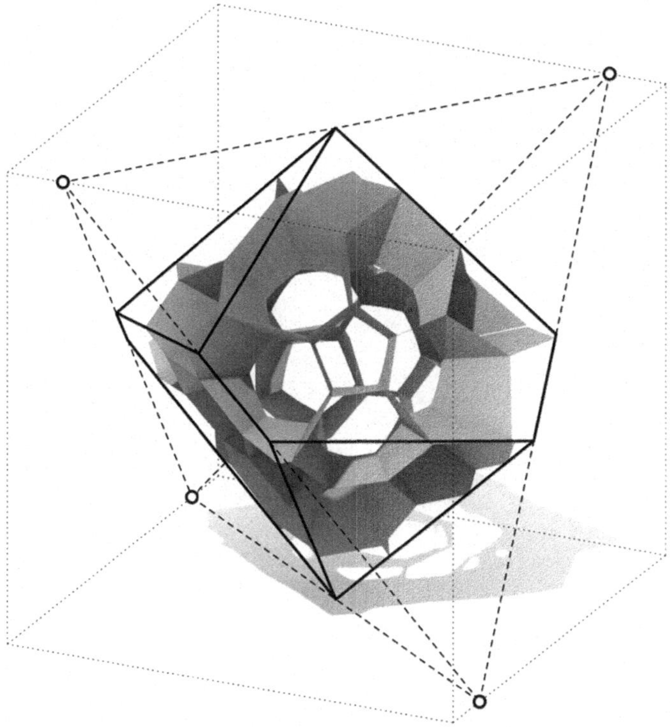

Fig. 1 Parametric design model

Fig. 2 Generation of Sparkler variations

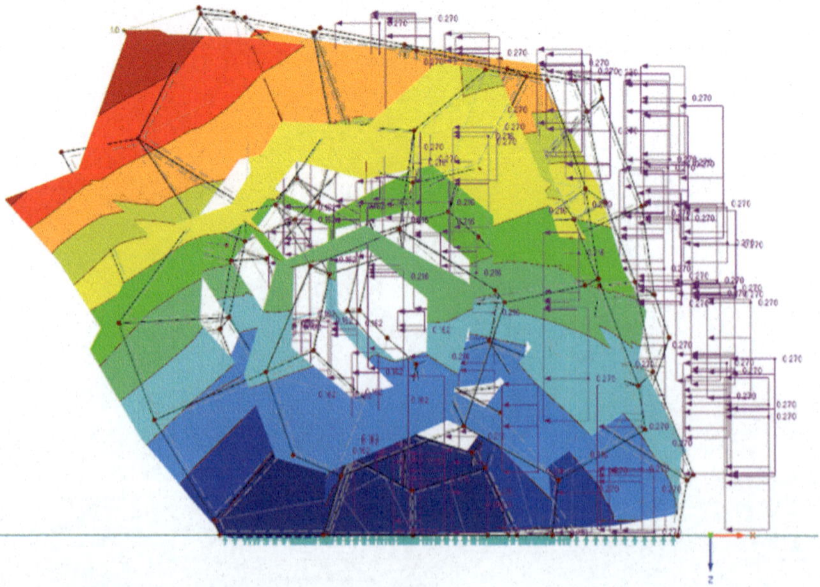

Fig. 3 Structural analysis. Wind load performance

Fig. 4 Lasercut scale model 1:10

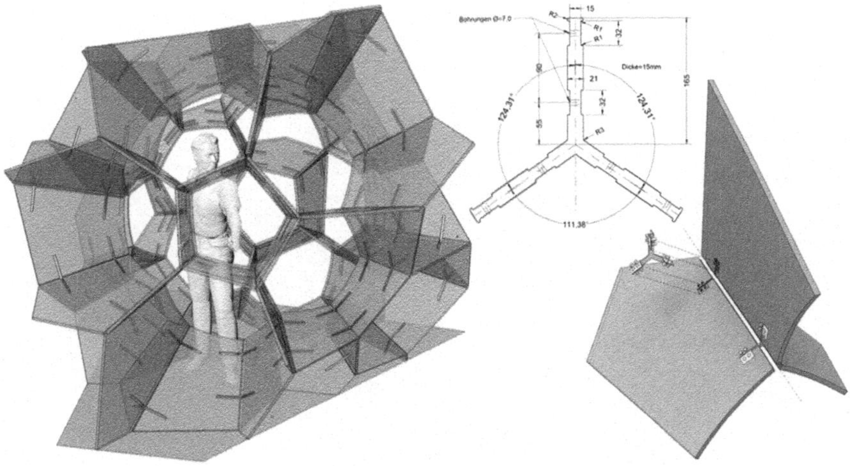

Fig. 5 Final parametric model, including material thickness and connection detail

Fig. 6 Assembly of the Sparkler pavilion

Fig. 7 Sparkler on the campus Emilie in Detmold

Fig. 8 Sparkler. View from outside and inside

Karamba

Clemens Preisinger, Marco Hemmerling, David Lemberski, Manuel Sotomayor, Rodrigo Velasco and Patric Günther

Master Computational Design and Construction (MCDC)
East-Westphalia University of Applied Sciences, Detmold, Germany (2013)

The design for the Karamba chair is based on a structural optimization process using Rhinoceros, Grasshopper, Galapagos, and Karamba as main software tools. The computational process involved the definition, analysis, and optimization steps in an iterative way, based on a system of variable constraints (parameters). This approach has been chosen intentionally to generate a performative design—in contrast to a design process, where the idea for a certain shape or appearance is driven by formal, visual, or purely aesthetic reasons. Starting with the definition of a closed curve for the initial outline of the seating element, the volume was generated by simply extruding the profile. After the definition of this volumetric container, an internal structure, based on Voronoi-like cellular configuration, was generated through the positioning of a point-cloud inside the previously defined container. In order to avoid random procedures such as 3D point population, a 3D boundary for points was defined whilst depending on structural analyses for its particular positioning. For that, an offset of the original curve was generated and re-parameterized to get a 0–1 margin in which any number of points could be located. Three-dimensional positioning

M. Hemmerling (✉)
Cologne University of Applied Sciences, Cologne, Germany
e-mail: marco.hemmerling@th-koeln.de

C. Preisinger · M. Hemmerling · D. Lemberski · M. Sotomayor · R. Velasco · P. Günther
Master Computational Design and Construction (MCDC),
East-Westphalia University of Applied Sciences, Detmold, Germany
e-mail: cpreisinger@bollinger-grohmann.at

D. Lemberski
e-mail: david.lemberski@hs-owl.de

C. Preisinger
Bollinger Grohmann, Vienna, Austria

M. Hemmerling and L. Cocchiarella (eds.), *Informed Architecture*,
DOI 10.1007/978-3-319-53135-9_18

capacity was added by moving the points on the curve space in a certain direction within the dimensional constraints of the extrusion.

After the generation of such a design model, supports were defined automatically by finding nodes with Z values of 0. Loads were applied to the sitting planes and in directions perpendicular to them, implying the possibility of loads for the upper back part of the seating element. Starting with random relative 3D positions of the central cell points given by the above-mentioned boundary, the Karamba structural analysis engine was used to evaluate the total deformation of the structure and went through an optimization process using Galapagos Evolutionary solver until a stable solution was found. Since the actual definition of the cells implied the use of planar elements only, fabrication was easily implemented using digitally driven sheet cutting processes, in our case, laser cutting. A Visual Basic script to unroll the cells was employed, and a Grasshopper definition was then applied to folding lines to convert them into dots, thus allowing for an all cutting geometry (as opposed to cut and engrave) whilst adding the flexibility of being able to fold the piece in either direction, which would be convenient in case of possible errors in unrolling directions (Figs. 1, 2, 3, 4, 5, 6, 7, and 8).

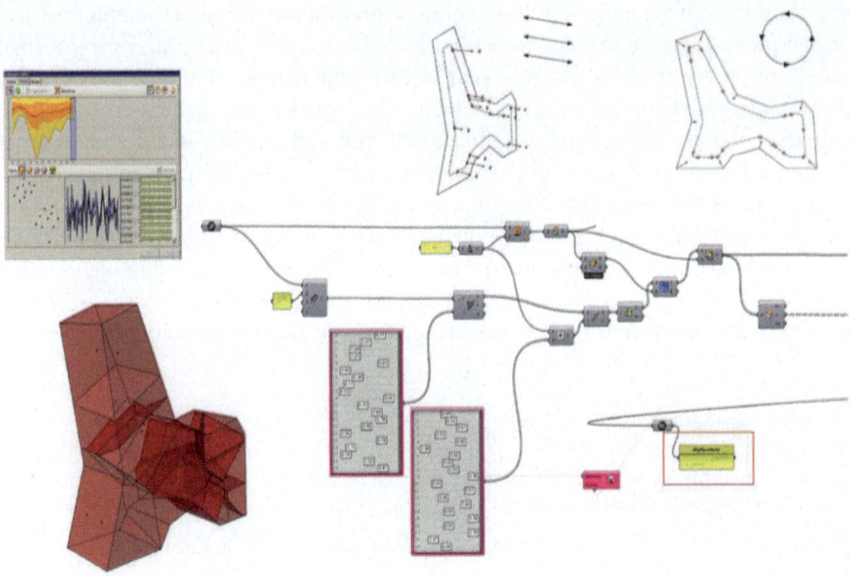

Fig. 1 Generation of the chair geometry

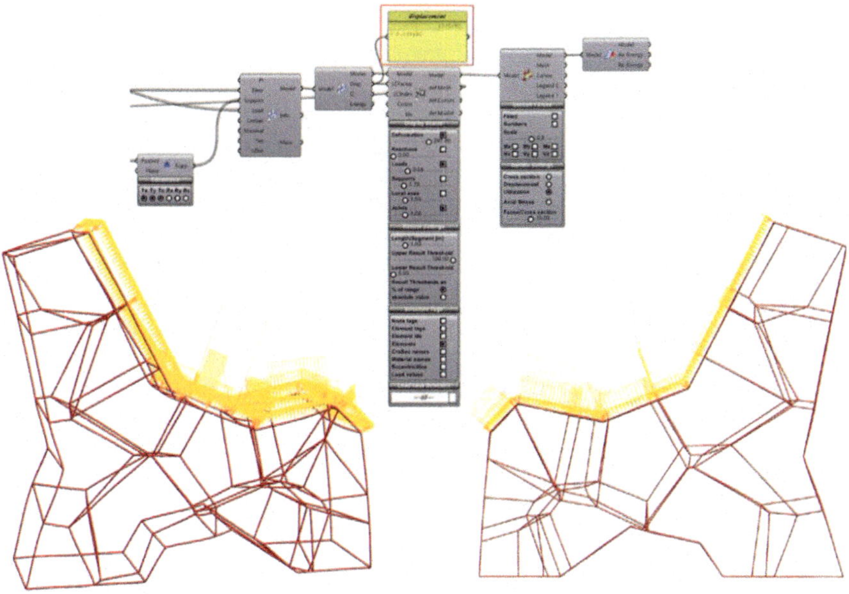

Fig. 2 Load analysis and first geometry optimization

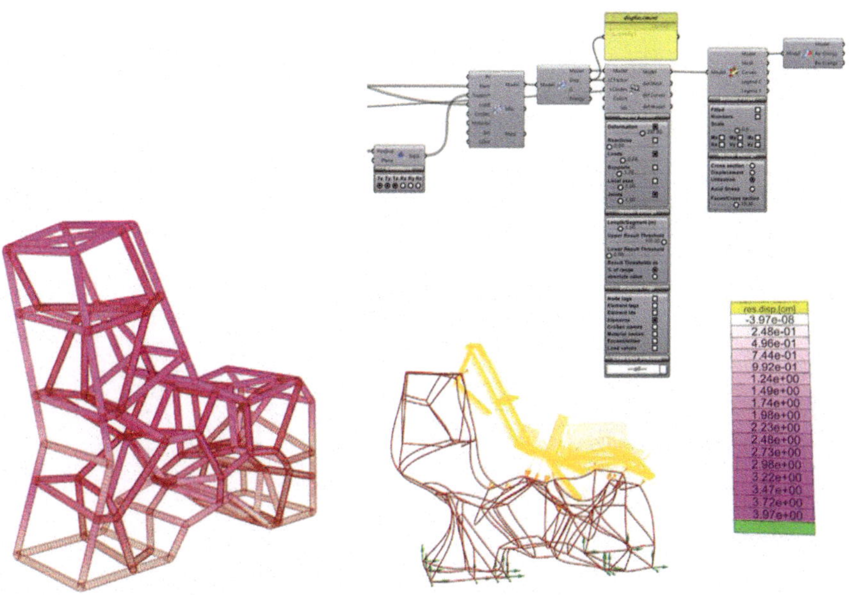

Fig. 3 Load analysis and second geometry optimization

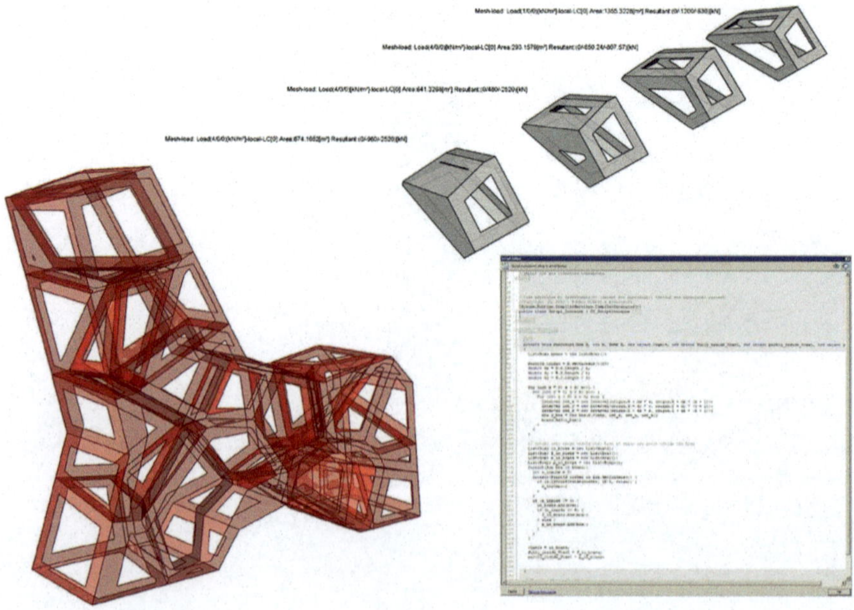

Fig. 4 Generation of the final cell-structure

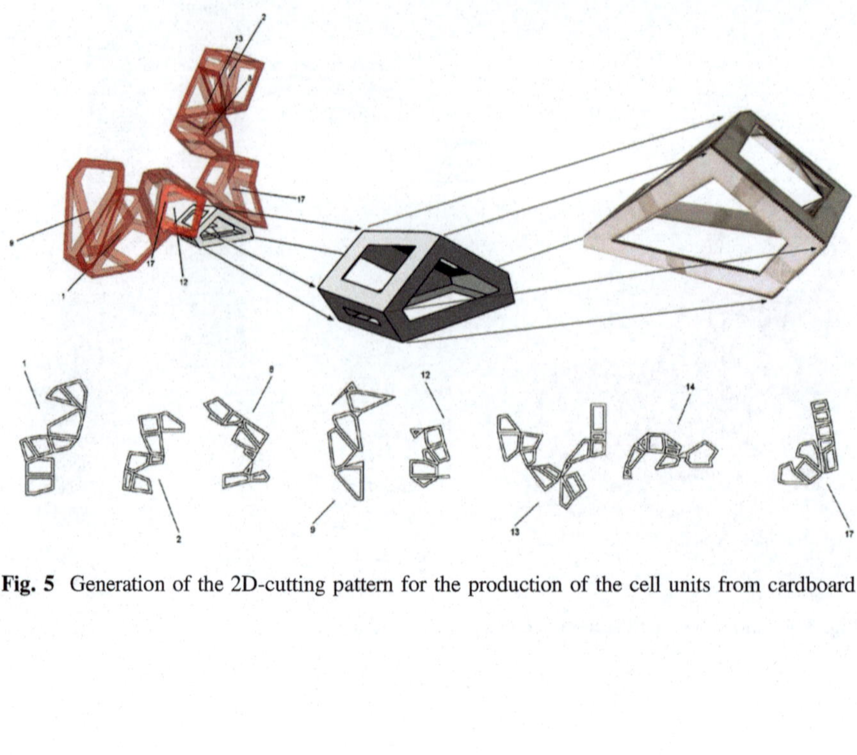

Fig. 5 Generation of the 2D-cutting pattern for the production of the cell units from cardboard

Fig. 6 Rendering and material study

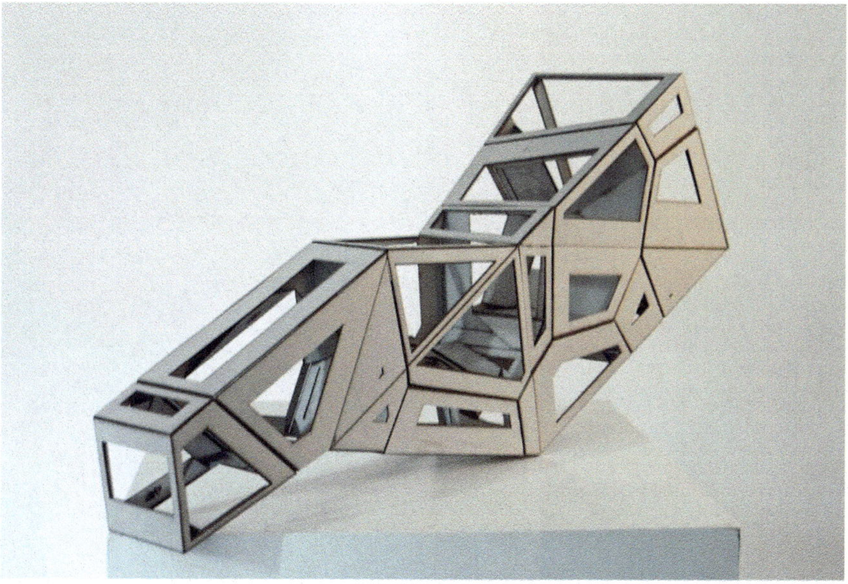

Fig. 7 Final cardboard model of a chair variation

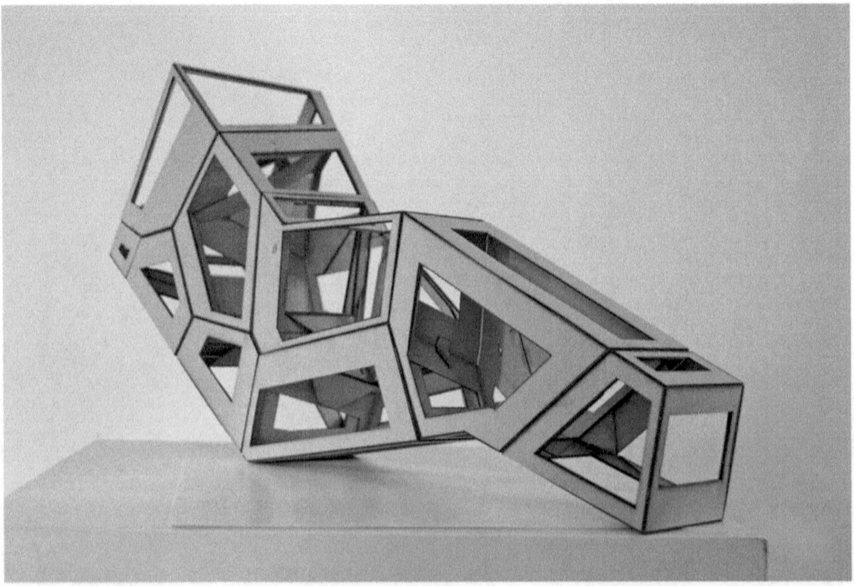

Fig. 8 Final cardboard model of a chair variation

Moebius

Christoph Schindler, Marco Hemmerling, Jens Böke, Manuel Sotomayor, Rodrigo Velasco, Patric Günther and Kaveh Allahdin

Master Computational Design and Construction (MCDC)
East-Westphalia University of Applied Sciences, Detmold, Germany (2013)

The project for the moebius bench was carried out as part of the digital fabrication module within the post-graduate Master program; Computational Design and Construction. The goal was not only to design a piece of furniture digitally in 3D but also to transfer it into a physical product in a 1:1 scale. For the production of the various parts, the panels were cut with a 5 axis CNC-Waterjet-Cutter.

The design process involved definition, analysis, and optimization steps in an iterative way as discussed below. It should be noted that as with any computational design process, the defined system was made out of variable constrains (parameters) as opposed to "designed" shapes. In the first step, a container was defined with an ellipse and a polygon which extruded along and also rotated around the edge of the ellipse. A seating area and a separator was defined after each 60 degree rotation of the polygon which, in the end, was chosen to be a triangle.

The topology of the material available (LG HiMacs composite panels) constrained the design itself to a construction made out of laminar pieces in two different directions; a horizontal plane serves as a guide and support for a set of

C. Schindler
Lucerne University of Applied Sciences and Arts, Lucerne, Switzerland
e-mail: christoph.schindler@hslu.ch

M. Hemmerling (✉) · J. Böke
Cologne University of Applied Sciences, Cologne, Germany
e-mail: marco.hemmerling@th-koeln.de

J. Böke
e-mail: jens.boeke@th-koeln.de

C. Schindler · M. Hemmerling · J. Böke · M. Sotomayor · R. Velasco · P. Günther · K. Allahdin
Master Computational Design and Construction (MCDC),
East-Westphalia University of Applied Sciences, Detmold, Germany

M. Hemmerling and L. Cocchiarella (eds.), *Informed Architecture*,
DOI 10.1007/978-3-319-53135-9_19

vertical pieces arrayed around the center of a bench that conformed the designed geometry itself. A maximum size of the given sheets forced a double layered construction of the horizontal framework to be used in order to ensure stability and the integrity of the whole.

A CNC cutting process automated the workflow from the designed pieces to the assembly process with a very high degree of accuracy. CNC represents an improvement on the accuracy of the produced elements, but also as an added difficulty in the assembling process. Higher accuracy means lower tolerances and an enhanced impact of inaccuracies in the assembling process. In order to avoid this difficulty, zip shapes were set to allow unique assembling positions of the neighbor pieces as well as a tongue and groove system for the vertical to horizontal assemblies. All the pieces were also labeled to ensure right positioning in the final model. The 5 axis WaterJet provided the chance for better interfacing on the built model, with the very first programmed geometry, by cutting the pieces in angled planes (Figs. 1, 2, 3, 4, 5, 6, 7, 8 and 9).

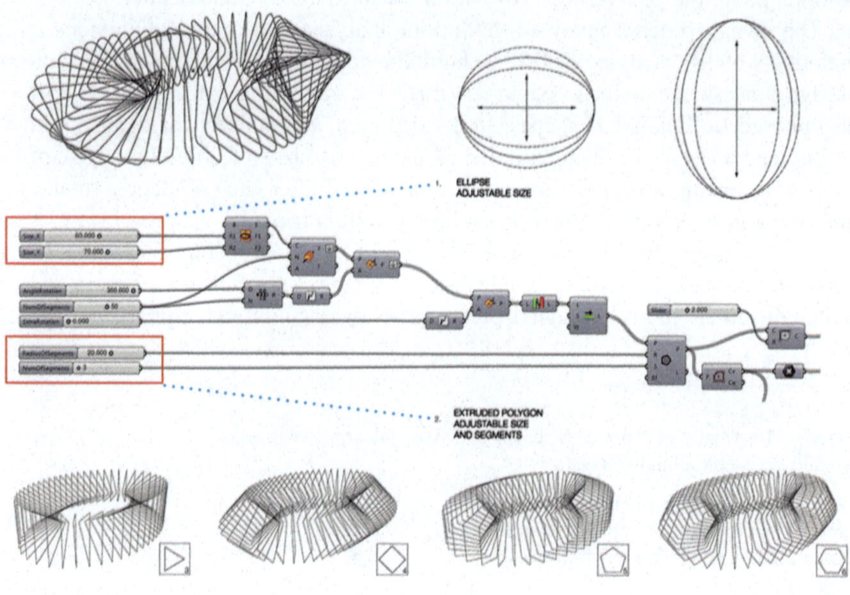

Fig. 1 Grasshopper definition of the moebius geometry

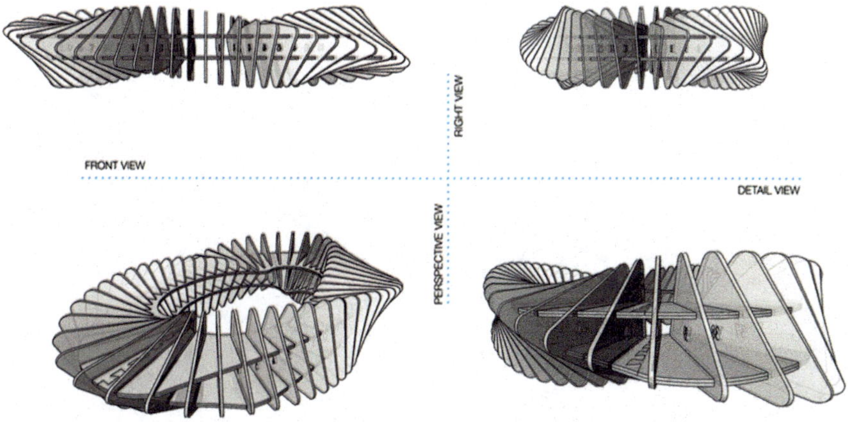

Fig. 2 Digital design model, including material thickness

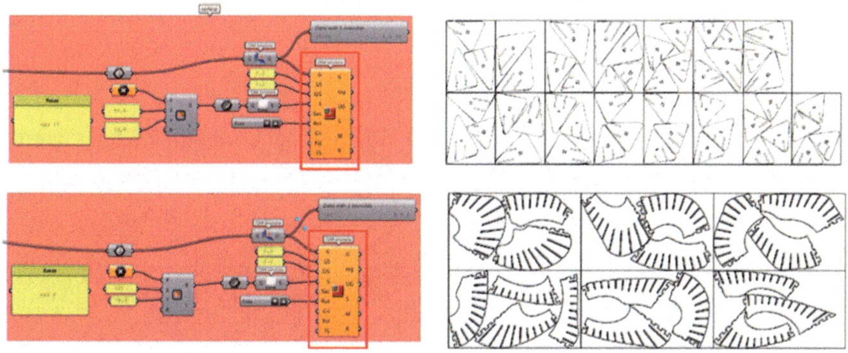

Fig. 3 Nesting of the various building parts for an optimized material use

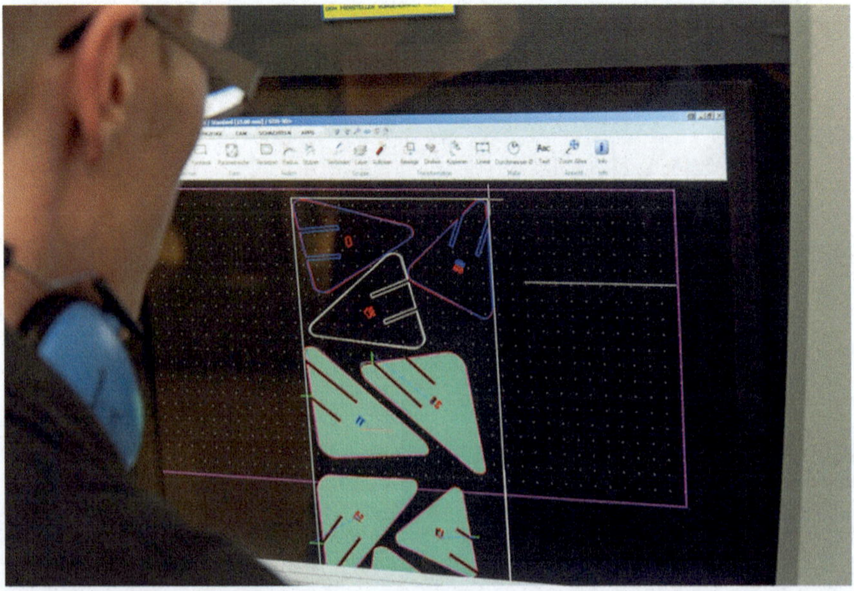

Fig. 4 Preparing the DXF-files for the WaterJet

Fig. 5 Cutting the composite panels with a WaterJet

Fig. 6 Taking out the building parts

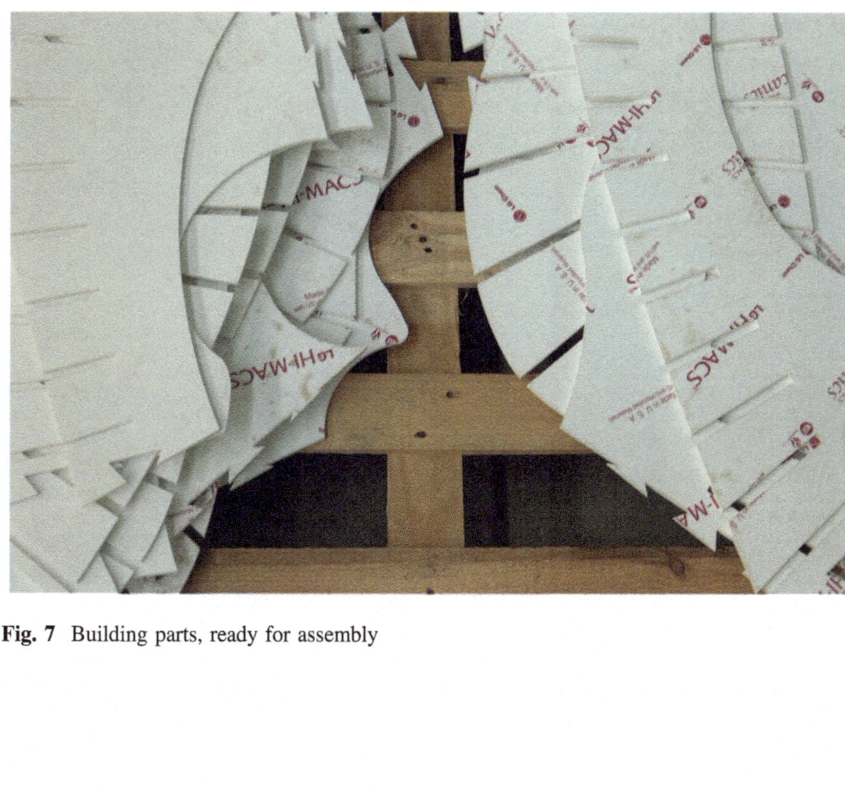

Fig. 7 Building parts, ready for assembly

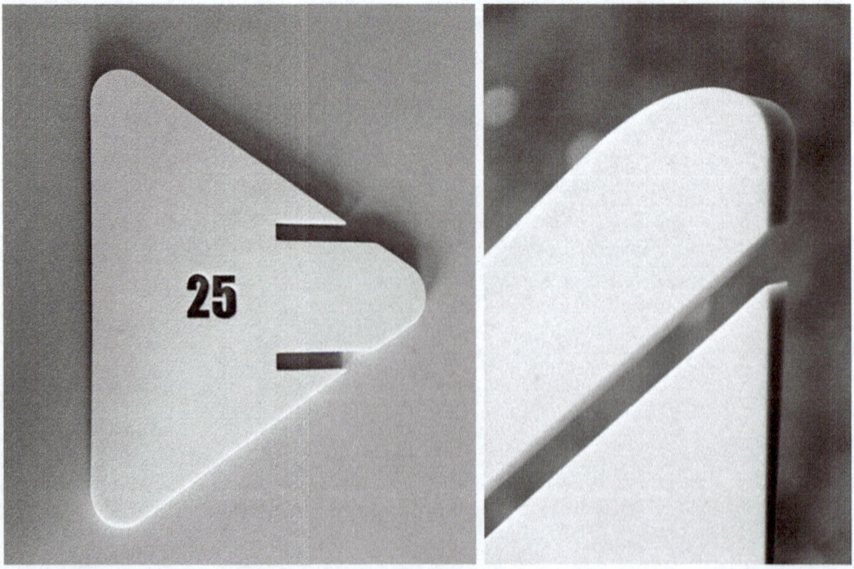

Fig. 8 Building part with label and detail of the intersection

Fig. 9 Moebius bench

Reciprocal Structure

Michael Knauß, Marco Hemmerling, David Lemberski, Guido Brand, Kaveh Allahdin, Eduard Benke, Larina Faßler, Lars Frenz, Tal Friedman, Jennifer Hilbert, Jan C. Kahre Heidemann, Linda Kirsten, Julian Lianarachchi, Vera Nolte, Henrik Steffen, Felix Stiboy, Ghaith Tish, Martin Trittin, Muralimanoj Varadharajan and Jirka Wolff

Master Computational Design and Construction (MCDC)
East-Westphalia University of Applied Sciences, Detmold, Germany (2014)

Throughout the last years, we have already seen promising advancements and applications in digital woodwork, such as CNC-technology and robotic manufacturing. Hence, the Master students took a closer look at the computational processes introduced to gain an in-depth understanding of the conditions and constraints as a starting point for their own design strategy. The four-day intensive program included both computational form-finding strategies in the design process as well as construction principles and digital fabrication technologies for the 1:1 realization of a reciprocal roof structure. Michael Knauß, founding partner of Zurich based office ROK and expert in digital fabrication, conducted the workshop during the *Detmolder Räume* (Detmold Spaces), a creative week of experimentation in interdisciplinary work on Campus Emilie.

M. Knauß · T. Friedman
Rippmann Oesterle Knauß GmbH (ROK), Zurich, Switzerland
e-mail: knauss@rok-office.com

M. Knauß · M. Hemmerling (✉) · D. Lemberski · G. Brand · K. Allahdin · E. Benke · L. Faßler · L. Frenz · J. Hilbert · J.C. Kahre Heidemann · L. Kirsten · J. Lianarachchi · V. Nolte · H. Steffen · F. Stiboy · G. Tish · M. Trittin · M. Varadharajan · J. Wolff
Master Computational Design and Construction (MCDC),
East-Westphalia University of Applied Sciences, Detmold, Germany
e-mail: marco.hemmerling@th-koeln.de

D. Lemberski
e-mail: david.lemberski@hs-owl.de

G. Brand
e-mail: guido.brand@hs-owl.de

M. Hemmerling
Computational Design in Architecture, Faculty of Architecture,
Cologne University of Applied Sciences, Cologne, Germany

M. Hemmerling and L. Cocchiarella (eds.), *Informed Architecture*,
DOI 10.1007/978-3-319-53135-9_20

The elective course aimed at the development of an innovative construction, based on self-supporting, reciprocal systems. The fundamental element of the system is the expanded node. A node in general is defined as a number of bar-elements meeting at or near one point within the system. Regarding fabrication and assembly of nodes, the complexity increases with the number of bars meeting at one point. The geometric expansion of the node reduces the complexity and allows a simple sequential joining process. This in turn creates a node with structural reciprocity, which has bending resistance and shear resistance, even though in its static system, all connections are pin-jointed. A minimum of 3 beams is required to create a reciprocal frame roof. As each beam supports the next in a reciprocal manner, no internal support structure is required. Only the outer end of each beam requires support which will normally be a post used for the wall or the foundation. Using this principle for a reciprocal frame, i.e. a roof structure where each beam both supports and is supported by other beams in the roof structure, it became the starting point for the project.

Each student team developed an individual spatial design based on these principles. Starting out from a 2D-grid, a 3D-spatial-frame was developed by using form-finding and simulation tools, e.g. Rhinoceros/Kangaroo or Sofistic. By applying a Python-Script to the spatial grid, the profiles were extruded along the axis-lines to generate the final geometry of the different bar-elements. In a first digital fabrication process, physical models in scale 1:10 were produce by laser-cutting the profiles from cardboard to verify the spatial concept, the geometric accuracy, and the assembly strategy. Finally, the building parts were produced from standard wooden profiles (40 × 60 mm) by cutting the length and the end-angles with a CNC-milling machine in combination with a blade saw. The assembly process itself was carried out in only a few hours due to the accuracy and easy principle for the connection of the bar-elements by sequential joining process (Figs. 1, 2, 3, 4, 5, 6, 7, 8, 9, 10, 11, 12, and 13).

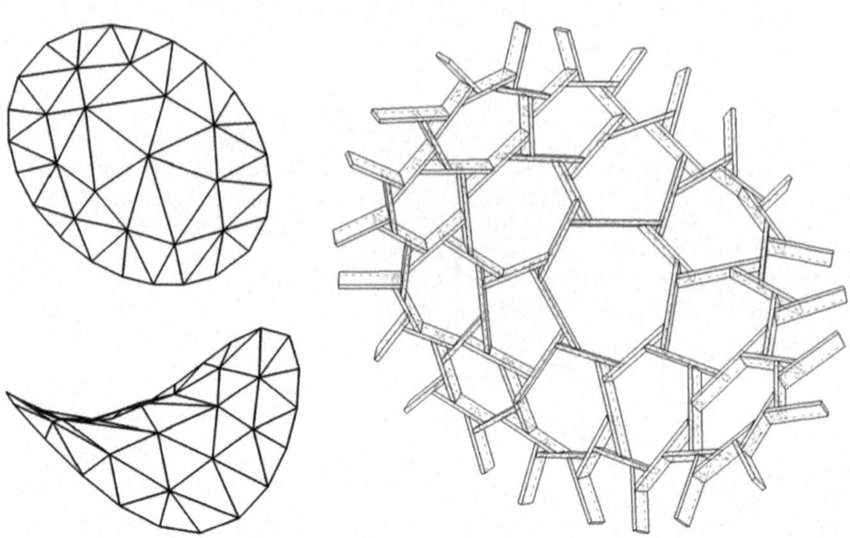

Fig. 1 Generation of the reciprocal structure, starting with a triangulated surface, which is subsequently transformed into hexagonal grid

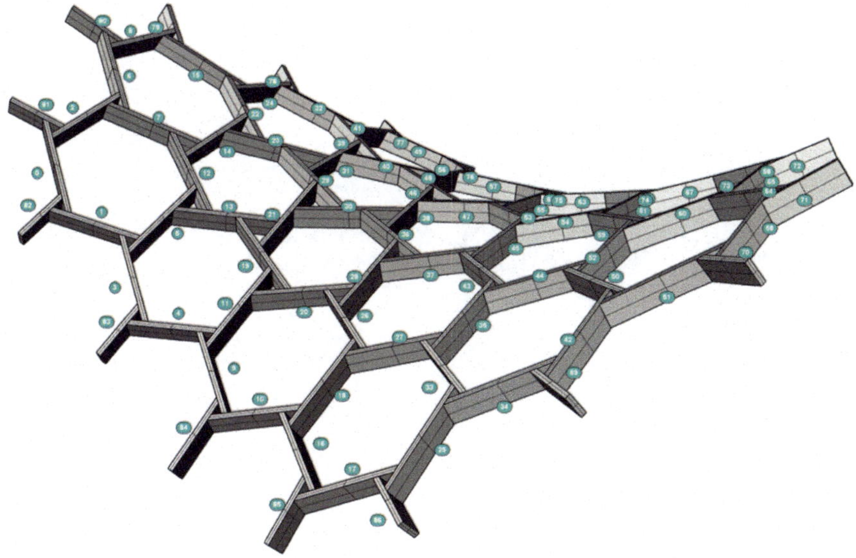

Fig. 2 3D-geometry of the reciprocal structure with labeling of the different members

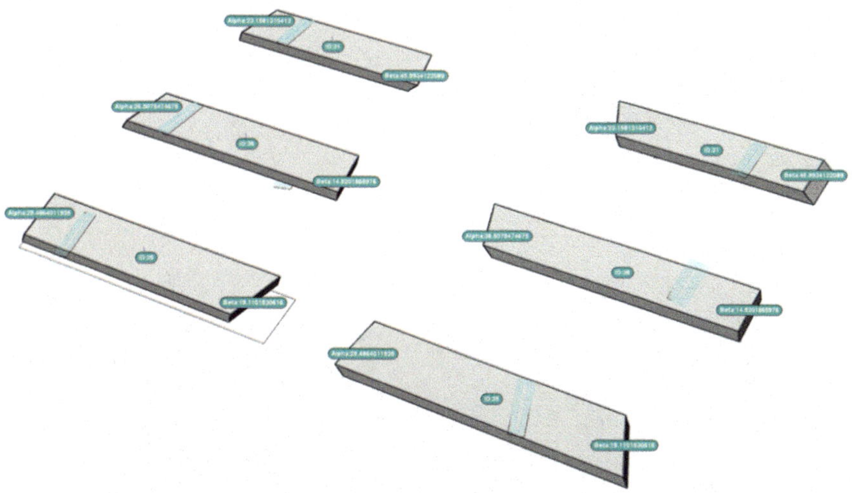

Fig. 3 Building members with label and position marks

Fig. 4 Computational modeling

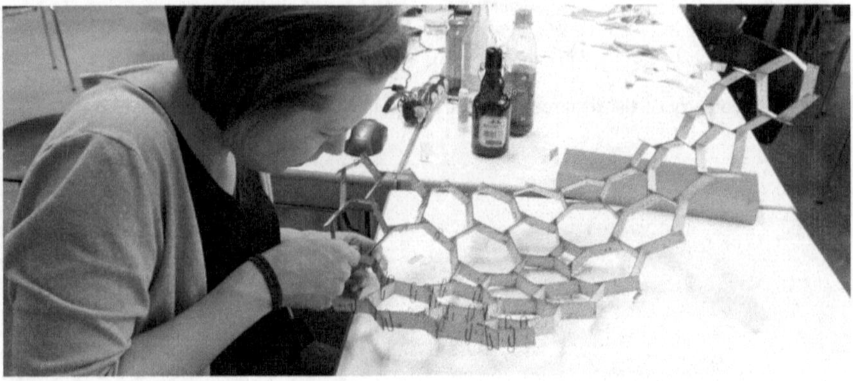

Fig. 5 Physical modeling

Fig. 6 Prototyping

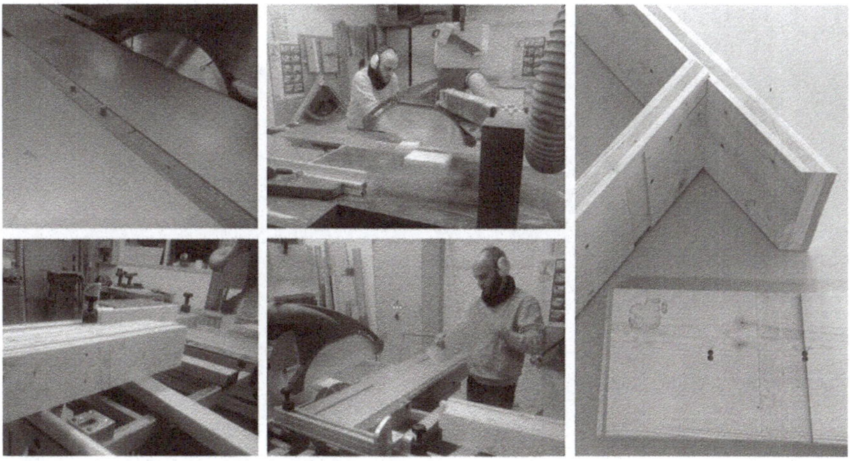

Fig. 7 Production of the building parts

Fig. 8 Testing the first reciprocal knot

Fig. 9 Assembly of the structure

Fig. 10 Positioning and testing the reciprocal structure

Fig. 11 Reciprocal structure on the Campus Emile in Detmold

Fig. 12 Reciprocal structure on the Campus Emile in Detmold

Fig. 13 Design proposal for a pavilion, based on the reciprocal structure

Origami

Marco Hemmerling, Jens-Uwe Schulz and Tal Friedman

Master Computational Design and Construction (MCDC)
East-Westphalia University of Applied Sciences, Detmold, Germany (2015–2016)

The Master thesis by Tal Friedman presents a novel approach to the construction of self-supporting thin-shell folded structures. It attempts to take the defining elements of Origami folding and implement them on an architectural scale, creating a full-scale pavilion folded out of composite aluminum boards which are parametrically designed to fold into shape. Using these principles, a design was made to approximate the growth of a flower which demonstrates an efficient thin shell self-supporting mechanism. A basic folding variable module was created which can be parametrically manipulated in order to form a seamless pattern. The method includes a workflow ranging from initial design to an FEA approach developed especially for the project and optimization algorithms for fabrication.

Classical Origami is based on the idea of an idealized zero thickness surface which is approximated by paper. However, in order to upscale this type of design methodology, a material thickness must be taken into account. In the process of understanding the nature of folds, various paper models were designed and built, starting off with thin paper and increasing material thickness gradually until a parametric paradigm was created that could be used to accommodate the design of 4 mm sheets. After the investigation of classical folding techniques, a method for the design and assembly of rigid composite aluminum boards was developed throughout the project. Algorithms were written in order to calculate and optimize the structure for various fabrication restraints. The project is made to resemble

M. Hemmerling (✉)
Cologne University of Applied Sciences, Cologne, Germany
e-mail: marco.hemmerling@th-koeln.de

M. Hemmerling · J.-U. Schulz · T. Friedman
Master Computational Design and Construction (MCDC),
East-Westphalia University of Applied Sciences, Detmold, Germany
e-mail: jens-uwe.schulz@hs-owl.de

M. Hemmerling and L. Cocchiarella (eds.), *Informed Architecture*,
DOI 10.1007/978-3-319-53135-9_21

Origami not only in its structural and aesthetically values but also in its fabrication method, which is actually folding the sheets rather than individual panels as is being done today in architecture. Each flower is made out of four unique sections which are folded into position. Each folded piece contains around 12-20 interconnected surfaces. This technique saves fabrication time and building tolerance. This project is a proof of concept model concluding that folded structures can be fabricated to full scale while maintaining a self-supported stiffness and stability. Reducing the need for supporting structural systems leads the way to new kinds of true thin shell lightweight structures made from just one material (monocoque). This method may have future applications in various fields, including, architecture, deployable structure, shading systems, and many more.

The structure draws its stability from the rigidity of the surfaces and by locking its fold angles in key locations. Hence, it is possible to reach a statically stable structure. A mathematical model was appraised in order to prognosticate the encastre points. For structural analysis, a computational model was analyzed using the Finite Element method. Due to the complex nature of the composite creased plates, a unique approach was developed on top of the basic formulations of thin shell elements in order to assess the differing thickness rather than treating it as a simple thin shell element. This requires a non-linear analysis which can take into account accumulating material fatigue (Figs. 1, 2, 3, 4, 5, 6, 7, 8, 9, and 10).

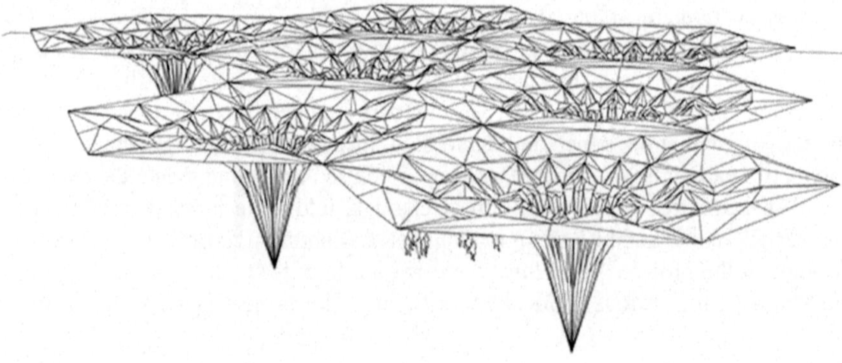

Fig. 1 First design sketch of the Origami pavilion

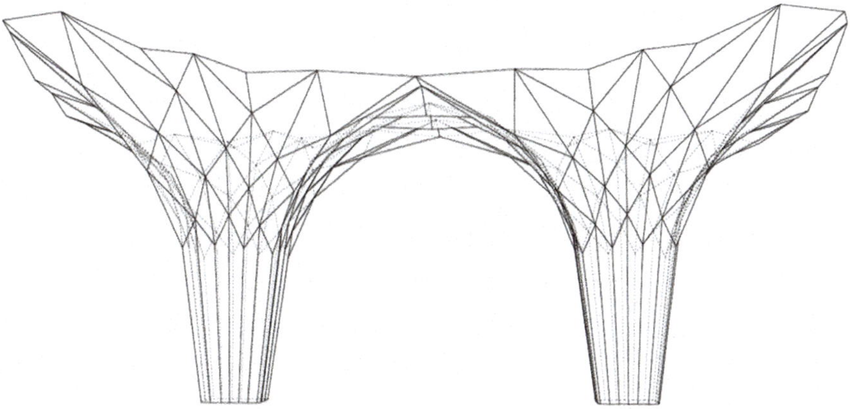

Fig. 2 Geometry-optimized Origami structure (*front view*)

Fig. 3 Geometry-optimized Origami structure (*top view*, *left*) and selection of a part for further investigation (*right*)

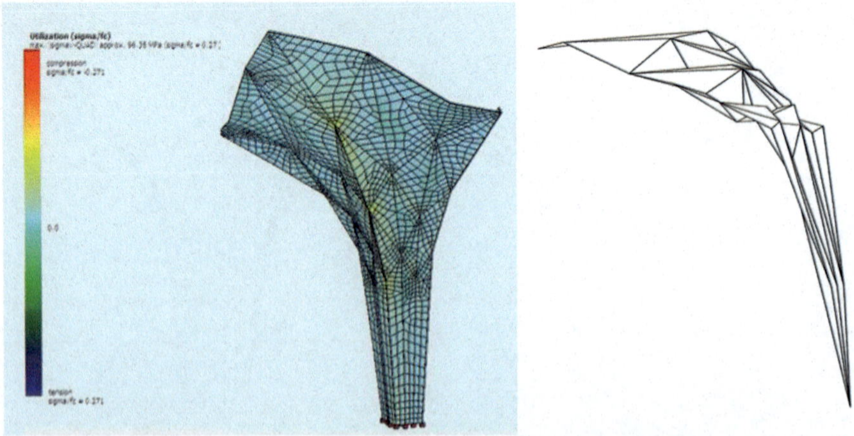

Fig. 4 Structural analysis (*left*) and optimized geometry of the selected building part (*right*)

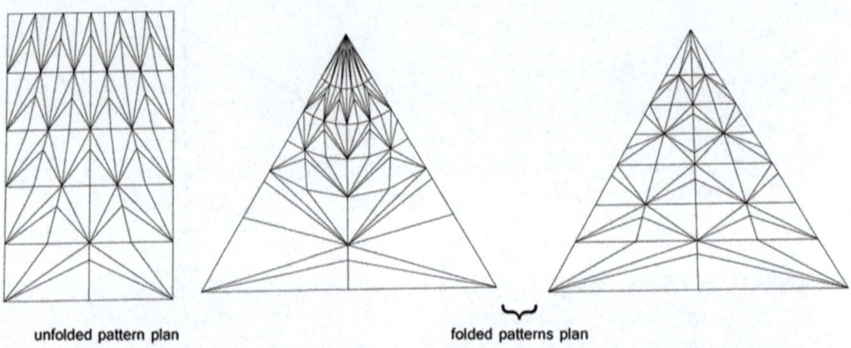

Fig. 5 Variation of the folding pattern

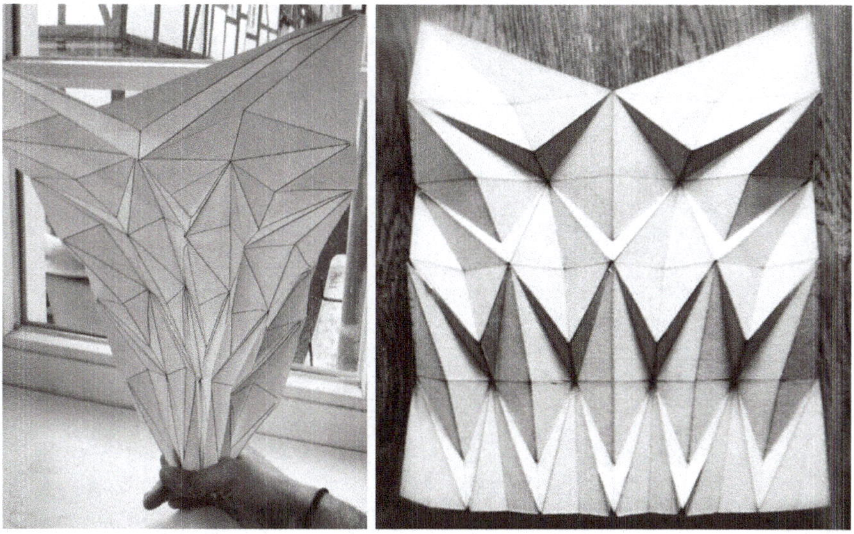

Fig. 6 Physical prototyping of the principle in paper and cardboard

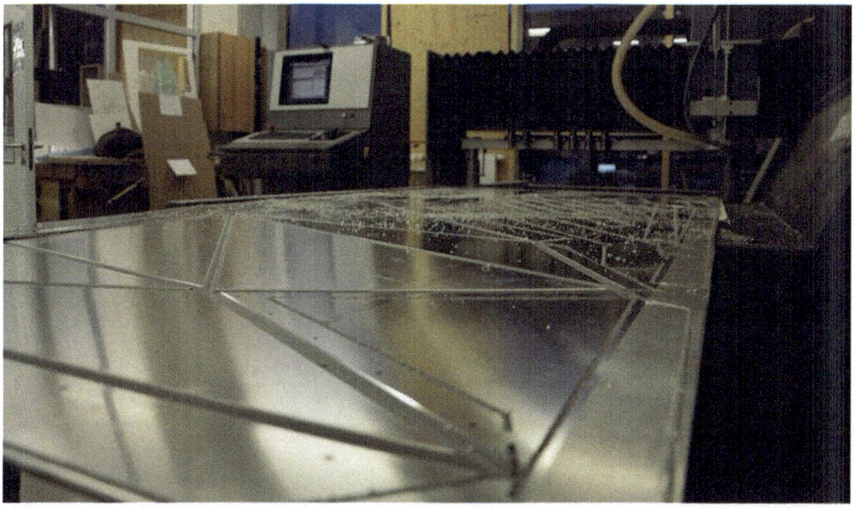

Fig. 7 CNC-milling the folding pattern into ACM boards

Fig. 8 Building the 1:1 mock-up

Fig. 9 Origami pavilion on the Campus Emilie in Detmold

Fig. 10 Detail of the Origami pavilion on the Campus Emilie in Detmold

(Un)Folding Space

Marco Hemmerling, Carlo de Falco, Alessio Mazzucchi, Giancarlo Gastaldin, Diego Oberti, Kristiyan Topchiev, Dominika Bugajska, Krzysztof Zinger, Atapaka Mohana Krishna, Sara Hajifathaliha, Shabarish Sathyanarayan and Shreyaa Jaya

Department of Architecture and Urban Studies School of Architecture, Urban Planning and Construction Engineering Politecnico di Milano, Italy (2017)

Folding strategies have been explored in architecture as a method to generate complex spatial and structural concepts from simple planar surfaces. The folding mechanism of a rigid material may seem as trivial as folding a sheet of paper. The striking elegance of models folded from paper appear more than just visually interesting. A simple crease along a line or a curve dictates a determined movement for both faces. The basic origami folds include valley and mountain folds, pleats, reverse folds, squash folds, and sinks. The number of basic folds is small, but they can be combined in a variety of ways to make intricate designs. Moreover, they inherit positive structural effects and allow for the creation of multiple pieces from one folded sheet, which may save material, fabrication time, and building costs. Hence, the ability to create a structural form from thin, flat materials have made folded structures an ideal candidate for lightweight deployable structures in architecture and engineering.

Within the framework of an interdisciplinary Master studio, led jointly by the departments of Architecture and Mathematics, the students developed, against this background, advanced spatial concepts, based on fairly simple mathematical definitions. The form-finding process started with the research of basic folding

M. Hemmerling (✉)
Computational Design in Architecture, Faculty of Architecture,
Cologne University of Applied Sciences, Cologne, Germany
e-mail: marco.hemmerling@th-koeln.de

C. de Falco
Department of Mathematics, Politecnico di Milano, Milan, Italy
e-mail: carlo.defalco@polimi.it

M. Hemmerling · C. de Falco · A. Mazzucchi · G. Gastaldin · D. Oberti · K. Topchiev · D. Bugajska · K. Zinger · A.M. Krishna · S. Hajifathaliha · S. Sathyanarayan · S. Jaya
Department of Architecture and Urban Studies, School of Architecture,
Urban Planning and Construction Engineering, Politecnico di Milano, Milan, Italy

M. Hemmerling and L. Cocchiarella (eds.), *Informed Architecture*,
DOI 10.1007/978-3-319-53135-9_22

principles, both in computational and physical modelling. The individual projects were developed in a bottom-up approach that allowed for the integration of findings throughout the design process.

The **Dot & Line** project started out with a research on translational and rotational surfaces. Therefore, a straight line was related to a central point by applying various sequences of rotational and translational operation. The resulting 3D models were examined against the constraint of developable surfaces in order to be able to produce an accordingly physical model of the resulting geometry from a flat sheet of paper. The finally chosen circular loop surface was not only a developable surface but also proved to be foldable without cutting an edge of the entire pattern. Hence, the kinetic property is inscribed in the continuity of the surface in such a way that it can adapt to different heights without changing its central diameter. In order to secure this feature in the computational parametric model, the Pythagoras theorem was applied to the rotation of the generating lines around the central point. After testing the principle with a series of paper models a final mock-up was CNC-milled from ACM-boards (diameter: 200 m) as a proof of the concept (Figs. 1, 2, 3, 4, 5, 6, and 7).

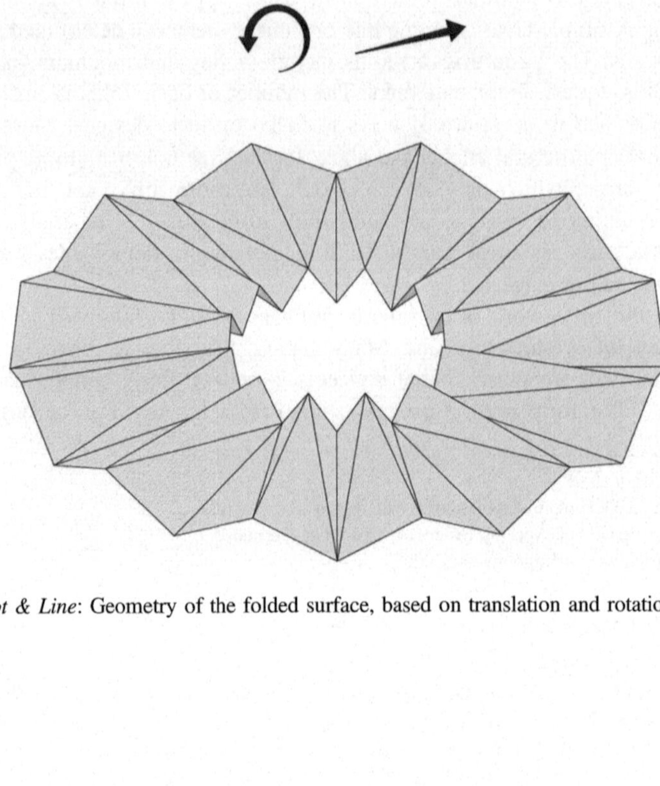

Fig. 1 *Dot & Line*: Geometry of the folded surface, based on translation and rotation

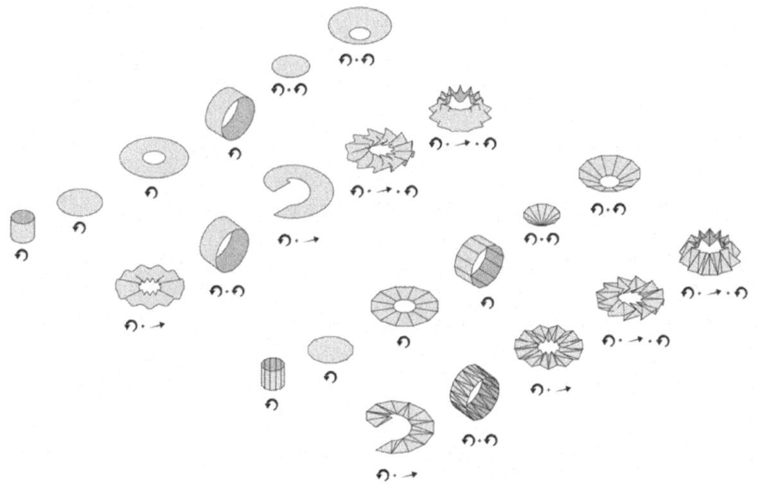

Fig. 2 *Dot & Line*: Variations of folded surfaces, based on translation and rotation

Fig. 3 *Dot & Line*: Physical prototypes from paper to verify the principle

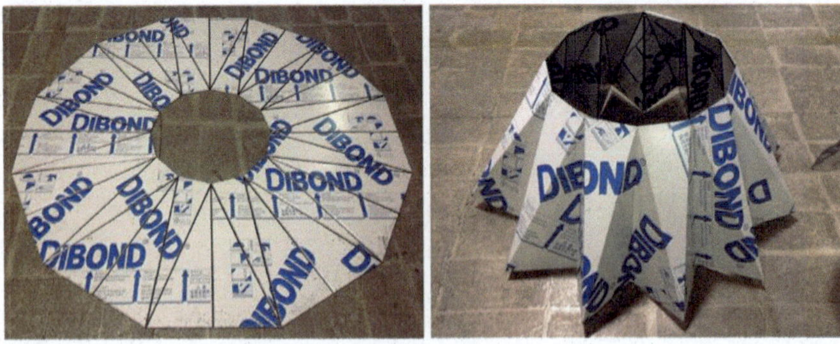

Fig. 4 *Dot & Line*: Cutting pattern (*left*) and folded structure from ACM board (*right*)

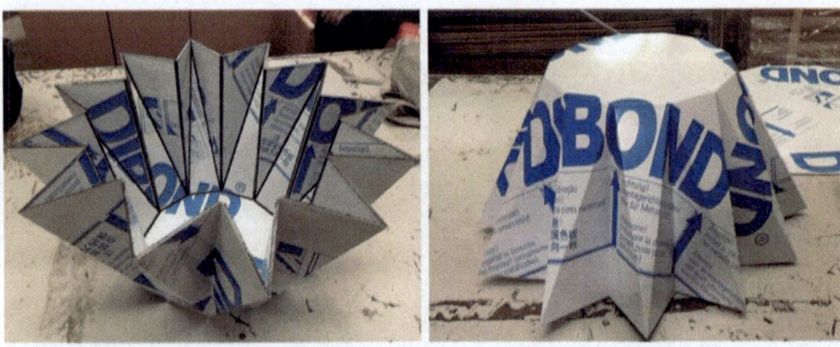

Fig. 5 *Dot & Line*: Variations of the folded structure from ACM board

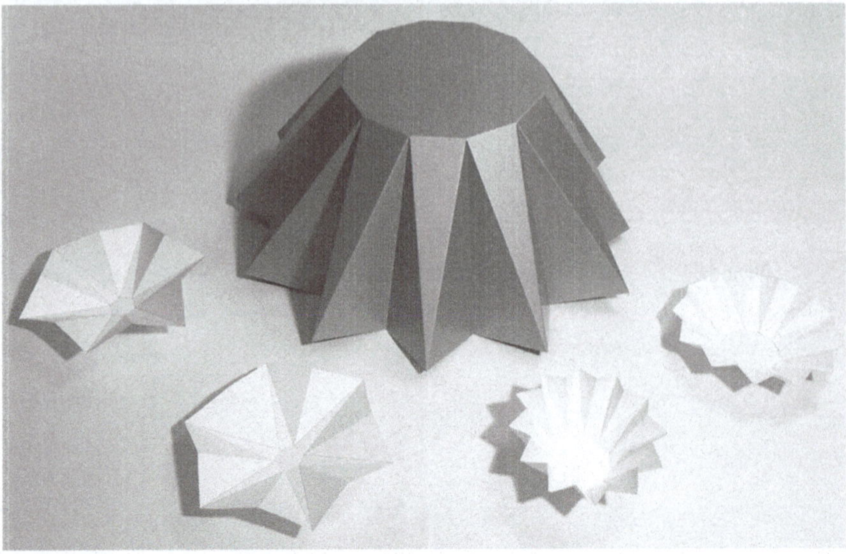

Fig. 6 *Dot & Line*: Prototypes from paper and final ACM-model

Fig. 7 *Dot & Line*: Final ACM-model

The interest in kinetic principles that can be found in folding was the driving force for the research of the project **Movement Trajectories**. Starting out again from a simple definition of the principle, the group developed a first computational model by defining a rotational movement around a fixed central point, which consequently resulted in a spherical surface. Gradually, the degrees of freedom and the number of points and axis's were increased in order to generate and control more complex movements. In a final transformation step, the folding control methods were adapted to simple mechanisms by defining two fixed base point, two points with determined trajectories and two points with trajectories defined by intersections with surfaces. As a result from this research, a series of simple mechanisms and systems with movable base points were computed, such as (semi)circular foldable surfaces and umbrella structures (Figs. 8, 9, 10, and 11).

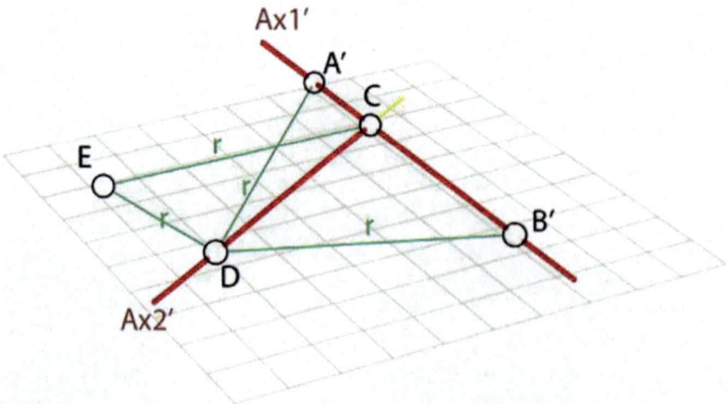

Fig. 8 *Movement trajectories*: definition of the basic geometry

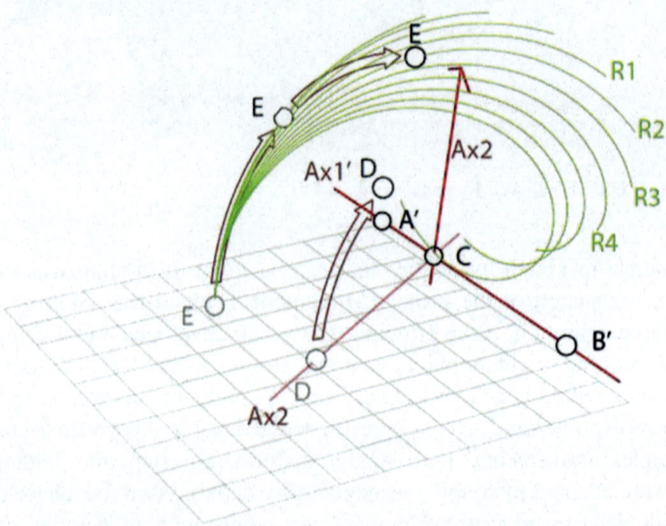

Fig. 9 *Movement trajectories*: generation of the movement trajectories

Fig. 10 *Movement trajectories*: variations of folding principles, based on the previously defined computational model

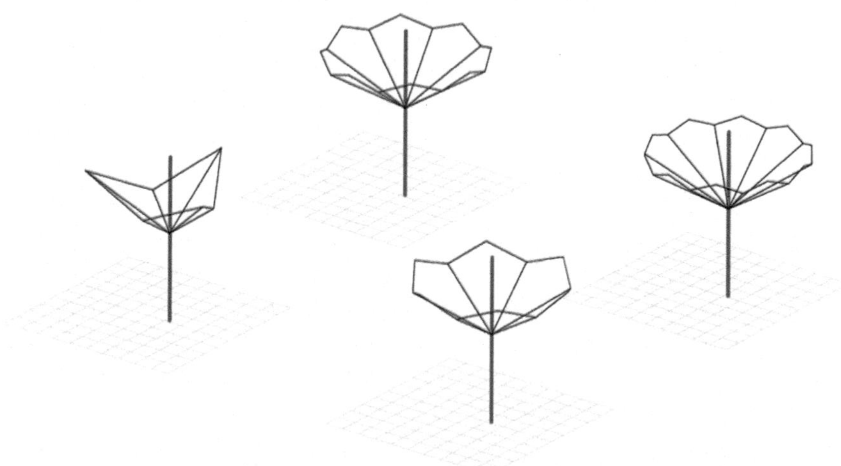

Fig. 11 *Movement trajectories:*: examples of an umbrella structure

A simple cube served as a starting point for the research **Solid Transformations**. In a first step, the computational design model was generated by dividing the edges of the cube by n points. The corresponding division points were then connected to generate a rotated square pattern on each face of the cube. The so defined squares were finally extruded and intersected to a unified solid by applying a Boolean operation. Based on this computational parametric model, a series of variation exploring the form generation principle was carried out. The research was concluded by unrolling the 3D surface to a 2D cutting pattern to produce a range of physical models in paper and finally from CNC-milled ACM boards in a bigger scale, as described in the first project (Figs. 12, 13, 14, 15, and 16).

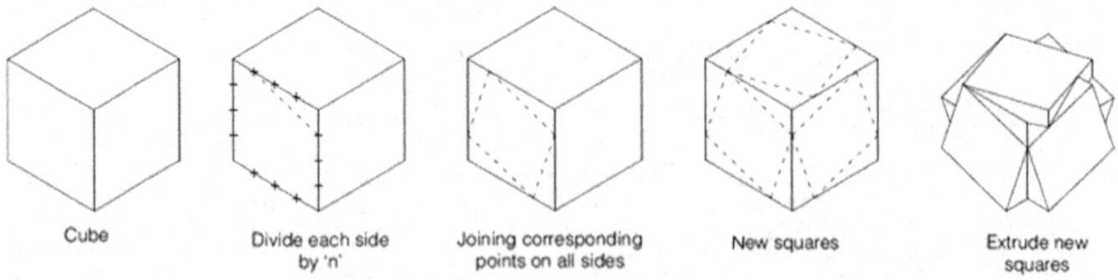

Fig. 12 *Solid transformations*: design principle for the generation of the folded structure

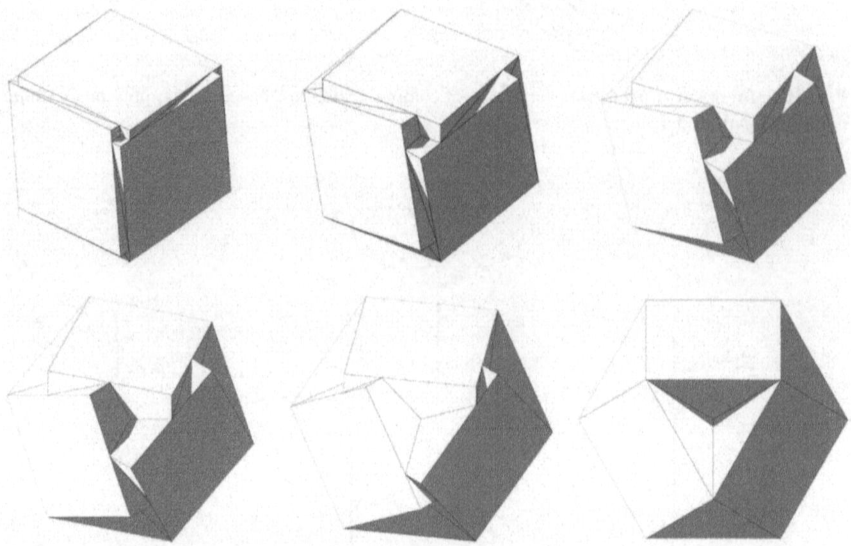

Fig. 13 *Solid transformations*: variations of the design principle

Fig. 14 *Solid transformations*: physical prototype from paper

Fig. 15 *Solid transformations*: assembly of the ACM-model

Fig. 16 *Solid transformations*: scale model from paper and final ACM-model

Printed by Printforce, the Netherlands